CARTOGRAPHY
PAST, PRESENT AND FUTURE

CARTOGRAPHY
PAST, PRESENT AND FUTURE

A *Festschrift* for F. J. Ormeling

Edited by

D. W. RHIND

Vice-President ICA

and

D. R. F. TAYLOR

President ICA

Published on behalf of the
INTERNATIONAL CARTOGRAPHIC ASSOCIATION
by
ELSEVIER APPLIED SCIENCE PUBLISHERS
LONDON and NEW YORK

ELSEVIER SCIENCE PUBLISHERS LTD
Crown House, Linton Road, Barking, Essex IG11 8JU, England

Sole Distributor in the USA and Canada
ELSEVIER SCIENCE PUBLISHING CO., INC.
655 Avenue of the Americas, New York, NY 10010, USA

WITH 3 TABLES AND 17 ILLUSTRATIONS

© 1989 ELSEVIER SCIENCE PUBLISHERS LTD

British Library Cataloguing in Publication Data

Cartography: past, present and future.
1. Cartography
I. Ormeling, F. J. (Ferdinand J.)
II. Rhind, David III. Taylor, D. R. F.
526

ISBN 1-85166-336-3

Library of Congress Cataloging-in-Publication Data

Cartography past, present, and future: a *festschrift* for F. J. Ormeling/
edited by D. W. Rhind and D. R. F. Taylor.
    p.   cm.
ISBN 1 85166 336 3
1. Cartography. 2. Ormeling, F. J. (Ferdinand Jan)
3. Cartographers—Netherlands—Biography.   I. Rhind, David
William
II. Taylor, D. R. F. (David Ruxton Fraser), 1937–
III. Ormeling, F. J. (Ferdinand Jan) IV. International Cartographic
Association.
GA108.C37 1989
526—dc19   89-1297
                                                              CIP

Phototypeset and Printed in Great Britain by Page Bros. (Norwich) Ltd., Norwich

PREFACE

This book is primarily a *Festschrift* for a remarkable individual: Fer Ormeling. The details of his career are outlined by his long-time colleagues Olof Hedbom and Rolf Böhme in the first chapter of this book. They describe Prof. Ormeling's substantial contributions to a number of organisations and causes. For us, however, the most important of these is the International Cartographic Association, the 'umbrella body' of world map-making. His involvement with the infant ICA and his successive adoption of the roles of Secretary/Treasurer and President span the whole lifetime of this now flourishing international scientific organisation. Appropriately, therefore, he has already written the history of the first 25 years of the Association. We have both known him for something like 18 years; in that time, we have worked, argued and laughed with him. We are delighted to have been able to mark his contributions to ICA in this way.

The idea for the book was born in January 1987 during a half-day excursion to Mussoorie, a hill town in northern India. The ICA Executive Committee, including Prof. Ormeling who was then the immediate past-President, was holding its six-monthly meeting in Dehra Dun at the invitation of General Agarwal (the then Surveyor General of India). Observing the majesty of the snow-capped Himalayas had the effect of dissolving—for a time—immediate problems and refocussing us on broader issues. We resolved to create a volume which would represent at least a sample of the 'state-of-the-art' of world cartography; thus, this *Festschrift*, unlike many, is a highly directed one. Contributors were, it is true, selected from amongst those with long and valuable service to ICA, often as Presidents, Vice-Presidents, organisers of conferences or as chairmen or chairwomen of Commissions. Much more important still, however, they were selected because they could and were prepared to provide a topical, challenging chapter which contributed to a carefully planned volume. We and the ICA owe them a debt of gratitude.

The book is, then, composed of two main sections after the opening chapter by Hedbom and Böhme on Prof. Ormeling. The first component is a section on the state of cartography as a whole in different countries; the second is a thematic or 'cross-cutting' view in which some of the major issues or developments in cartography are examined in turn.

Clearly, no examination of the 'state-of-the-art' can be exhaustive so we have taken a sample representing the multiplicity of countries in the ICA. No significance should be attached to the sequence in which countries are described; this simply follows the alphabetical order of the names of the authors. Each author was given the freedom to select what he felt to be the most important national issues to raise. We begin with Prof. Adalemo and Dr Balogun, who describe the situation in Nigeria, and they are followed by General Agarwal who summarises the equivalent situation in India. Doctor Baranowski sets out what is happening, and has happened, in map-making in Poland and Prof. Bertin and Dr Duch-Gary complement this with accounts of developments in France and Mexico. Finally, Prof. Freitag summarises the state of cartography in the Federal Republic of Germany, and Prof. Hu Yuju and Mr Fei Lifan complete the section with a paper on the People's Republic of China.

If much of this first section is factual, personal interpretations are also unavoidable. Thus the accounts differ in form and style, reflecting both the situation in that country and the emphases chosen by the contributor. The second section, however, introduces a still greater level of personal synthesis and views; fortunately, all of the authors are well known throughout the cartographic world for their scholarship. The section is begun by a pair of papers by Prof. Arthur Robinson and Dr Árpád Papp-Váry; trained in very different cultures, they examine the nature of cartography as either an art or a science—or both. Doctor Helen Wallis then summarises the status of one of the most popular yet demanding forms of scholarship in map-making: historical cartography.

It is a particular pleasure to include a chapter in this section by Prof. Ferjan Ormeling, son of Fer. His inclusion, however, reflects no form of nepotism: he is very widely known and respected for his work in cartographic education. His paper is followed by contributions from each of us. These reflect preoccupations over many years. Thus D.R.F.T. writes on the cartography in the Developing World whilst D. R. (greatly aided by a part-time PhD student, Ruth Blatchford) attempts to offer a view on what, after two decades or more of computer cartography, an ideal mapping system should be able to do for the cartographer. The whole book is closed by a chapter by Dr Joel Morrison. At the time when this book was initiated, Joel was President of the ICA and travelling even more widely than his industrious predecessors. This, together with his long-term involvement with a variety of international and national organisations, make his review of the revolution in cartography in the 1980s, mostly due to the advent of the computer as a routine tool, a most authoritative one.

Our operating principle in compiling this book has been to change only as much of the submitted text as was vital to permit the understanding by the reader of sometimes complex concepts or subtle hints in the individual papers. We have, in addition, imposed a degree of standardisation on the use of English (c.f. American) English! In all of this, of course, contributors are responsible for the statements they make, unless we have inadvertently introduced error—in which case, we are responsible for it.

We have already indicated our debt to the contributors. In addition, however, many other individuals have contributed to the final product, notably many colleagues, relatives, friends and our publishers. We trust that they will think the end result makes it all worthwhile!

David Rhind and D. R. Fraser Taylor

CONTENTS

II Thematic issues in cartography

LIST OF CONTRIBUTORS

PROF. DR ISAAC A. ADALEMO

Department of Geography, University of Lagos, Lagos, Nigeria
Isaac A. Adalemo is Professor of Geography and Deputy Vice-Chancellor, University of Lagos. He graduated AB, AM, PhD from the University of Michigan, Ann Arbor, Michigan, USA, where he studied under Professors Waldo Tobler and George Kish, and is currently President of the Nigerian Cartographic Association and Vice-Chairman of the Nigerian National Committee of the International Cartographic Association. He was instrumental in developing the Laboratory for Cartography and Remote Sensing as a unit of the Department of Geography at the University of Lagos, and is keenly interested in the development of cartographic education at both university and pre-university levels.

MAJOR-GENERAL G. C. AGARWAL

37 Curzon Road, Dehra Dun, Utter Pradesh, India
At the time his paper was written, General Agarwal was Surveyor General of India and a Vice-President of ICA. He advised the government of India on all surveying and mapping matters and was responsible for the work of the Survey's 20 000 employees.
Graduating with a Bachelor of Engineering degree from Roorkee University, he went on to obtain a degree in photogrammetry (MSc, PhE) from the International Training Centre for Aerial Surveys (ITC), then based in Delft in the Netherlands. He joined the Survey in 1954,

became Head of the Photogrammetry Division of the Indian Photo-interpretation Institute in 1956 and was appointed as a Director of the Survey of India in 1970 and Director of Research Development in 1974, finally taking over as Surveyor General in 1981.

General Agarwal has published widely on mapping matters and has been honoured by the Prime Minister of India for his work on the Hindi–English Dictionary of Technical Terms in Cartography, and by the Chief Minister of Andhra Pradesh for work on the Planning Atlas of that state. He has been Chairman of the Indian National Cartographic Association, a member of the ICA Commission III and member of many governing boards in India and overseas. In addition, he has made many contributions to the International Society of Photogrammetry and Remote Sensing. After his retirement from the Survey of India, General Agarwal became Vice-Chancellor of Maharshi Dayanand University, Rohtak, (Haryana).

DR OLAYINKA Y. BALOGUN

Department of Geography, University of Lagos, Lagos, Nigeria
Yinka Balogun is an Associate Professor of Geography (specialising in Cartography) in the University of Lagos. He is a geography graduate of the University of Ibadan and cartography graduate of the University of Wisconsin, Madison, USA. He is a former Director of the University of Lagos' Laboratory for Cartography and Remote Sensing, the First Vice-President of the Nigerian Cartographic Association and Editor of the *Nigerian Cartographer*. His research interests include psychophysical studies of map symbols, the development of mapping in Nigeria and he is currently looking at the development of a Geographical Information System for Nigeria. A member of a working group for the production of the *Atlas of the Federal Republic of Nigeria*, he is also a member of the ICA Commission on Training and Education in cartography.

DR MAREK BARANOWSKI

Ul. Symfonii 4 m. 16, 02-786 Warsaw, Poland
Marek Baranowski graduated from Warsaw University with an MA in Geography in 1973. He received his doctorate in 1980 for work in Computer Assisted Cartography from the Institute of Geodesy and

Cartography in Warsaw. From 1973 to 1984 he worked as a researcher in the Geodetical Cartographic Data Processing Institute, becoming Director in 1988 in Warsaw. In 1988 he moved to the Institute of Geodesy and Cartography as Head of the Geographical Information Systems Laboratory. He continues to carry out research in Computer Assisted Cartography as well as GIS. A frequent contributor to ICA meetings, he has also lectured in several countries such as Iraq and Algeria. He lectures in the University of Warsaw and in the Technical University of Warsaw.

PROF. DR JACQUES BERTIN

53 rue Dareau, 75014 Paris, France
Prior to retirement, Prof. Bertin was Director of Studies at the School of Advanced Studies in the Social Sciences (EHESS) which is part of the École Pratique des Hautes Études (EPHE, VIe Section), Paris, and was Director of the Laboratoire de Graphique, Paris. He studied geography and cartography at the University of Paris where he received his doctorate. He founded the Cartography Laboratory of EPHE, later transforming it into the Laboratory of Graphics. His major publications include the seminal work *Sémiologie Graphique* (1967), *La Graphique et la Traitment Graphique de l'Information*, *Théorie Matricielle de la Graphique et de la Cartographie* and *Traitments Mathématiques et Traitments Graphiques—Différence et Complémentarité.*

RUTH BLATCHFORD

c/o Department of Geography, Birkbeck College, University of London, 7–15 Gresse Street, London W1P 1PA, UK
Ruth Blatchford (née Hartley) gained a degree in geography at the University of Durham, UK, and then worked on the data base and mapping aspects of the CORINE (Coordination of Environmental Information for Europe) project in Birkbeck College. Her studies for a doctorate are supervised by Prof. Rhind and are centred on the design of graphics for both ephemeral and hard-copy mapping, especially for road mapping. She is now employed by Pinpoint Ltd.

ROLF BÖHME

Solmser Straße 18, D-6361, Niddatal 1, FRG
Dipl. Ing. Rolf Böhme was educated in Geodesy and Geography at the universities of Dresden and Hannover, graduating in 1949. From 1952, he was employed at the Institüt für Angewandte Geodäsie (Institute for Applied Geodesy) in Frankfurt am Main in the Federal Republic of Germany. In 1972 he became head of its Department of Cartography, with special responsibility for small-scale mapping, and held this post until his retirement in 1982. He wrote numerous publications, both national and international (especially for *The International Yearbook of Cartography*). His responsibilities also included being Secretary of the German Cartographic Society in the 1970s and serving as a member of the UN Group of Experts on Geographical Names. Rolf Böhme was a Vice-President of the ICA from 1976 until 1984 and was awarded an Honorary Fellowship of the Association at the General Assembly held in Perth, Western Australia.

DR NESTOR DUCH-GARY

Patriotismo 711-PH, Col. Mixcoac, 03910 Mexico, D.F.
Doctor Duch-Gary is Director General of Geography in the National Institute for Statistics, Geography and Data Processing, Secretaria de Programacion Y Presupuesta, Mexico. He has also received the distinction of being appointed as Secretaria de la Presidencia (1976).

He received a degree in Economics from the Instituto Politecnico Nacional in 1968. He served as head or assistant head of several departments before being appointed as Director General of Geography in the National Institute in 1983. He has also served as Professor of Economics and Professor of Statistical Methods in the University of Mexico.

Doctor Duch-Gary has written many professional publications on socio-economic and statistical aspects of cartography and on automated cartography. He has served as ICA Vice-President from 1984 and was in charge of the organisation for the 1987 ICA General Assembly and International Conference in Morelia, Mexico.

FEI LIFAN

Wuhan Technical University of Survey and Mapping (WTUSM), 39 Loyu Road, Wuhan 430070, People's Republic of China
Mr Fei is a lecturer and Director of the Research Section on New Technology, in the Department of Cartography in the Wuhan Technical University of Surveying and Mapping: Born in 1949, he graduated from WTUSM in 1977 and got his MS degree in 1982. Since then, Mr Fei has been carrying out research and teaching in the field of Computer-Assisted Cartography; this work has led to the publication of a series of papers and text books in China and elsewhere.

PROF. DR ULRICH FREITAG

Department of Cartography, Free University of Berlin, Arno-Holz-Straße 12, D-1000 Berlin 41, Germany
Professor Freitag studied Geography in the Humboldt University, East Berlin, and Cartography in the Academy of Construction in West Berlin. He lectured in the Justis-Liebig University in Gießen between 1962 and 1966, also gaining his PhD at the same time. This was followed by two years lecturing in Geography at the University of Ife in Nigeria. Returning for four years as a Senior Lecturer (then Professor) to his former university, he then worked from 1973 until 1976 as Visiting Professor at the Chulalongkorn University in Bangkok and was Technical Advisor to the Royal Thai Survey Department. In 1977, he became a full Professor and took the Chair of Cartography in the Department of Geography at the Free University of Berlin.

Amongst his many responsibilities have been the co-editorship of *Kartographische Nachrichten* for two periods; he has served as Vice President and (currently) as President of the German Cartographic Society and Member or Chairman of various Working Groups and Commissions of the ICA.

OLOF HEDBOM

Flottbrovagen 16, 11264 Stockholm, Sweden
Olof Hedbom, born in Stockholm in 1920, was instructed by his Geography teacher at the age of 16 to produce the basis of a wall map

series on the Swedish shipping industry. The maps were completed and exhibited at the opening of the Navy Museum in 1937; from that time and for a period of over 50 years, Olof Hedbom made his living as a commercial cartographer.

His university training was under Prof. Ahlmann, a President of the International Geographical Union, and covered geography, geology, photogrammetry and psychology. Whilst studying, he was also employed at the Esselte Map Division, headed by Dr Carl Mannerfelt. At 25, Hedbom became chief editor and later manager of the Kartografiska Institutet of Esselte and was responsible for the cartography in the Swedish National Atlas between 1953 and 1971. From 1975 to 1985, he was director of Libor Kartor. He created the plan for the second version of the National Atlas of Sweden, to be published—by Parliamentary decree—in 20 volumes.

At the same time, he occupied a number of honorary positions, such as President of the Swedish Society for Anthropology and Geography. His formal contribution to international mapping began in 1958 as a member of the IGU Commission on National and Regional Atlases (later a joint IGU/ICA Commission). He was a Vice-President of ICA 1972–76 and Secretary–Treasurer from 1976 to 1984, being elected Honorary Fellow in 1987.

PROF. HU YUJU

Wuhan Technical University of Survey and Mapping (WTUSM), 39 Loyu Road, Wuhan 430070, People's Republic of China
Professor Hu works in the Department of Cartography, Wuhan Technical University of Survey and Mapping. He is also the Editor-in-Chief of China's cartographic journal *Cartography*. Born in 1927, he graduated from Tongji University, Shanghai, in 1952 and lectured at that university until his appointment to Wuhan.

Hu Yuju has published many textbooks and professional papers on cartography, especially on map projections, mathematical cartography and cartographic education. He served as local organiser for the ICA Advanced Seminar on Cartographic Education held in Wuhan in 1986. In addition, Dr Hu has been an ICA Vice-President since 1984.

DR JOEL MORRISON

US Geological Survey, 516 National Center, Reston, Virginia 22092, USA
Doctor Morrison is Assistant Division Chief of Research in the National Mapping Division of the United States Geological Survey. Prior to his employment in the Survey, he was a Professor in the Department of Geography at the University of Wisconsin in Madison, USA, where he taught on the internationally known cartography programme. He served as Chairman of the Department and as Director of the Cartographic Services Laboratory.

Over the years, Dr Morrison has published papers in journals such as *The American Cartographer*, *Cartographica* and *The International Yearbook of Cartography*. He is co-author of *Elements of Cartography* and has served on many national and international committees on various aspects of mapping. Joel Morrison was Vice-President of ICA from 1980 until 1984, and President from 1984 to 1987, travelling widely in the service of ICA.

PROF. DR FERJAN ORMELING

Geografisch Instituut, Heidelberglaan 2, Postbus 80115, 3508 TC Utrecht, The Netherlands
As part-time atlas editor at Wolters–Noordhoff Publishing House, Ferjan Ormeling financed his training as a geo-cartographer at Groningen and Utrecht Universities from 1961. In 1969 he joined Cor Koeman's staff at Utrecht University where he became responsible for thematic cartography. He instigated cooperation with the other academic carto-graphic training facilities in The Netherlands, at Delft Technological University and at ITC, with which a joint excursion, fieldwork and lecture programme was established. In 1983, he took his doctor's degree with a thesis on minority toponyms on maps. After Koeman's departure in 1981, Ormeling was appointed as his successor.

In 1969 he joined the editors of The Netherlands' cartographic journal, now called *Kartografisch Tijdschrift*. In 1973 he became a member of the Board of The Netherlands Cartographic Society, for which he organised a number of national conferences (on tourist maps, three-dimensional representation, map use, map testing and on visual percep-tion). From 1972 onwards he has participated in the work of the ICA Commission on Cartographic Education.

DR ÁRPÁD PAPP-VÁRY

MÉM Országos Földügi és Térképészeti Hivital, Kossuth Lajos tér 11, H-1860 Budapest 55 Pf: 1, Hungary
Doctor Papp-Váry is Head of the Department of Geodesy and Cartography, National Office of Lands and Mapping of the Ministry of Agriculture and Food Industry, Budapest.

He was one of the chief editors of the *Planning-Economic Atlas Series of Hungary* (1974) and the new *National Atlas of Hungary* (1989), and was co-author of the book, *The Map as a Mirror of the Earth* (1983). He has been a Visiting Professor at the Chair of Cartography of the Eotvos Lorand University in Budapest since 1974, lecturing on atlases and the history of Hungarian surveys and cartography.

Doctor Papp-Váry has been an active member of the ICA since 1972. He has served as a member of the Commission on Cartographic Technology and Committee on Statutes and Bylaws. He has acted as Co-Chairman of the Standing Commission on Map Production Technology. He has served as both Secretary and President of the Hungarian National Committee of the ICA, and initiated the successful proposal to host the 14th ICA Conference in Hungary in 1989. Doctor Papp-Váry was elected as a Vice-President of ICA in 1987.

PROF. DR DAVID RHIND

Department of Geography, Birkbeck College, University of London, 7–15 Gresse Street, London W1P 1PA, UK
David Rhind is Professor of Geography and Head of the Economics, Geography and Statistics Resource Centre at Birkbeck College, University of London. Much of his research work has been in computer-assisted cartography and, since 1970, in Geographical Information Systems. His publications include the co-authored *People in Britain—A Census Atlas* and *An Atlas of EEC Affairs*, as well as over 100 papers and various reports.

He has served as advisor to various committees such as the House of Lords' Select Committee on Science and Technology and was a member of the UK government's Committee of Enquiry on the Handling of Geographical Information. Doctor Rhind has been a Vice-President of the ICA since 1984 and is currently Chairman of the Royal Society's Ordnance Survey Scientific Committee and Honorary Secretary of the Royal Geographical Society.

PROF. DR ARTHUR H. ROBINSON

101 Burr Oak Lane, Mount Horeb, Wisconsin 53572, USA
Arthur Robinson PhD, LittD, DSc(Hon), directed the Map Division of the Office of Strategic Services from 1941–45. He taught in the Department of Geography, University of Wisconsin-Madison until his retirement in 1980. At Wisconsin, he founded the University Cartographic Laboratory of which he was Director from 1966–73, being named Lawrence Martin Professor of Cartography in 1967.

He served as President, Association of American Geographers in 1964 and was President of the ICA from 1972–76, receiving the Mannerfelt Medal in 1981. His books and monographs include *The Look of Maps*, *Elements of Cartography* (in five editions), *The Nature of Maps*, *Fundamentals of Physical Geography*, *Early Thematic Mapping in the History of Cartography* and *Cartographic Innovations* (co-editor); his other publications include many research papers. In addition, he was the founding editor of *The American Cartographer* and Honorary Consultant in Cartography to the Library of Congress.

PROF. DR D. R. FRASER TAYLOR

Department of Geography, Faculty of Social Sciences, Loeb Building, Carleton University, Ottawa, Canada K1S 5B6
Doctor Taylor is Associate Dean (Academic) of the Faculty of Graduate Studies and Research at Carleton University, Ottawa, Canada. He is also Professor of Geography and International Affairs and has twice been President of the Canadian Cartographic Association.

His numerous publications on cartography include editing and contributing to the series of books on *Contemporary Cartography*, published by John Wiley and Sons, Chichester. Having been ICA Vice-President from 1984, he succeeded to the Presidency at the General Assembly held in Morelia in Mexico in 1987.

DR HELEN WALLIS

c/o The Map Library, British Library, Great Russell Street, London WC1B 3DG, UK
Doctor Wallis joined the then Map Room of the British Museum in 1951 as an Assistant Keeper; she was Deputy Keeper from 1967 and became

Map Librarian of the British Library (BL) when the library departments of the British Museum were transferred to the BL in 1973. She retired from paid employment in the BL in 1986 but retains active research interests therein.

Whilst at the BL, Dr Wallis organised numerous exhibitions, such as those on 'The American War of Independence', 'The Famous Voyage of Sir Francis Drake' and 'Raleigh and Roanoke'; many of these were subsequently transferred to other museums. She has authored many catalogues and learned papers and co-edited *Cartographical Innovations* with Arthur H. Robinson.

Helen Wallis was Chairman of the ICA Standing Commission for the History of Cartography between 1976 and 1987. In addition, she is President of the International Map Collectors Society and has been Chairman of the Society for Nautical Research since 1970. Her work has been recognised by such honours as the award of Membre d'Honneur de la Société de Géographie de Paris, a medal of the British Cartographic Society and the Order of the British Empire.

FERDINAND J. ORMELING: A BIOGRAPHY

Olof Hedbom and Rolf Böhme

Ferdinand J. Ormeling was born in Amsterdam on April 12th, 1912. He was married twice and has three children from his first marriage: Piet Hein, Ferjan (a contributor to this volume) and Ina. From his marriage in 1949 to Regina (Rini) Kamerbeek, he also has three children: Erik, Sonja and Roger.

He spent his childhood in Amsterdam and Hilversum and saw as a young boy the effects of World War I (1914–1918). He took his matriculation examination at the age of 18 in Hilversum High School. From the age of 18 to 23 he studied at the State University of Utrecht, specialising in history and geography. After passing his university examinations, he became a teacher for gymnasia in Hilversum and The Hague.

THE TROPICAL PERIOD

During World War II (1939–1945), he was in the Netherlands' army and after the capitulation of the Dutch, refusing to report to the Germans, he had to go 'underground'. In 1945 he was sent to the Netherlands' East Indies to restore the pre-war colonial situation. Having a geographical and cartographic background, he was taken out of the field troops and transferred to Batavia (now Jakarta) to serve with the new Geographical Institute (1947). This was then a branch of the 'Topografische Dienst', a part of the government mapping agency of the former Netherlands' Indies. This Institute was charged with various cartographic tasks and with the compilation of allied geographical studies based on aerial photography. The Institute also produced various terrain studies. On April 12, 1982 during the festive occasion of his retirement from ITC, Ormeling presented a copy of these intelligence reports to his 'former adversary and present friend, Ir. Asmoro Pranoto, from Indonesia'.

In 1949, Indonesia was declared an independent, sovereign state and most of the Dutch people had to leave the country. Some of them stayed, however – among them F. J. Ormeling. It required special personal qualities to work (especially doing field studies) in a country which had recently been ruled by the Dutch Army and, later, by the Japanese military forces – and which, moreover, was badly damaged during the war. Ormeling managed to be accepted as a working companion by people who shortly before had been the official enemies of his country. Thus, after Indonesia's liberation, he was appointed by the new Indonesian Government as Head of the Geographical Institute which then had 45 employees. The main tasks were to produce small-scale maps, to make geographical surveys and to organise the education and training of the personnel of the Indonesian Topographic Service. His many journeys and fieldwork made him well-acquainted both with the history and the human geography of Indonesia. In 1950, he had the opportunity to carry out research and fieldwork on the island of Timor, thanks to support of the first Indonesian Director of 'Jawatan Topografi', Soerjo Soemarno, a former student of the Technical University of Delft, Netherlands. During various longer visits over a period of three years, Ormeling studied the specific ethnic and socio-economic characteristics of the island, which he then summarised in a thesis for his Doctor's degree: 'The Timor Problem, A Geographical Interpretation of an Underdeveloped Island'. As a token of appreciation for the assistance he had received, he presented his thesis to the University of Indonesia which awarded him the title of Doctor in Social Sciences. The book was one of the first to focus on an underdeveloped region in a modern way. It quickly sold out and was then reprinted in a second edition – something rare for an academic book.

In 1955, his 10 year tropical period came to an end. It was a period which formed an important mental base for his coming life as a geographer, cartographer and administrator. He had learned to respect and understand people in developing areas and their problems.

THE ATLAS MAKER

Outstanding school atlases have been made in several countries since the 19th century. In the Netherlands, this atlas is the 'Bosatlas', published since 1877. In 1955, when the Ormelings returned home from Indonesia, the atlas publisher J.B. Wolters of Groningen was looking for a geographer-cartographer with experience in education to up-date and renovate their main product. Ormeling was appointed to do the job, which was to last for more than 20 years. Starting as an internal employee, he ended up as an external collaborator. He was the editor responsible for nine editions of the *Grote Bosatlas* and six editions of the *Kleene Bosatlas*, and during that time the atlases changed dramatically – not only in cartographic design but also in content. They were enlarged by the inclusion of hundreds of excellent thematic maps, produced in close cooperation with Dutch teachers, and were published always under strict financial and timetabling constraints.

THE ORGANISER

Together with a small group of Dutch cartographic enthusiasts – among them Romein and Koeman – Ormeling organised a new cartographic section of the Royal Netherlands Geographical Society in 1958. This section was gradually recognised by surveyors, photogrammetrists and geographers and formed a bridge both between cartographic institutes in the Netherlands and with colleagues in cartography abroad. Ormeling was Chairman of this section for ten years. In 1975, the section became the independent Dutch Cartographic Society. From 1967 to 1971, he was President of the Royal Netherlands Geographical Society, the first President after the merger of the various Dutch geographical associations.

Between 1967 and 1984, he represented the Netherlands and the ICA as a member of the United Nations' Group of Experts on Geographical Names. As the Chairman of the Working Group on Education, he organised training courses on toponomy in Indonesia (1982) and Morocco (1984).

THE PROFESSOR

In 1964, at the age of 52, he was appointed Professor of Economic Geography at the University of Amsterdam. There he established the

Economic Geography Institute which concentrated on economic-geographical research of Amsterdam. The 1960s were the period of 'democratisation' of higher education in the Netherlands and also of the advance of the quantitative revolution in geography. Professor Ormeling described the situation in his valedictory address 'Final Project' when he retired from ITC in 1982, as follows:

> 'One of the most obvious reasons for this development is what is called the 'quantitative revolution' – an inevitable companion of the positivist philosophy that invaded Geography in the 1950s, and which assailed the map as being subjective and descriptive, and replaced it by quantitative spatial analysis and model-building. In my inaugural address on economic geography in Amsterdam in 1964, I welcomed the quantifiers in the innocent anticipation that visual thinking by means of maps and symbolic thinking in formulae would complement and support each other. My maps, however, were swept away by the flood of quantitative methods and techniques.'

ITC

The first Prime Minister of the Netherlands after World War II was Professor Dr Jr W. Schermerhorn, who was also a famous photogrammetrist. He was a man with global views and became the creator of the International Training Center (now the International Institute for Aerospace Survey and Earth Sciences) or ITC. It was founded in Delft in 1951 as a contribution from the Netherlands to international co-operation in development. In 1971, the Institute moved to Enschede, close to the border with the Federal Republic of Germany. In the same year, Professor Ormeling was appointed to the Institute where he created and became Head of the Department of Cartography. It benefited from the ready availability of funds in the Netherlands in the 1970s, particularly for Third World purposes. Consequently, he managed to provide the Department with adequate modern equipment, including sophisticated computer hardware. At that time, his thoughts on the professional cartographic situation were given in his introductory essay entitled 'Turbulent Cartography'.

The aims of the ITC happened to coincide very closely with the aims of the International Cartographic Association – cartographic education, dissemination of the languages of maps, close relationships with the Third World, the publication of books on scientific and technical cartography, etc. Thus, during the term of his tenure of office at the Institute between 1971 and 1982, Professor Ormeling and his staff had around

Fig. 1. Professor Ormeling in academic dress waving goodbye to his students and colleagues on the occasion of his retirement from ITC, April 1982.

450 students from about 140 countries and regions. His personal concern for the students and his vivid and inspiring teaching are well remembered by those who had the advantage of being his students at ITC.

Like almost all ICA executives and Commission chairmen, Professor Ormeling also had 'normal employment' with a salary: ICA service has to be performed on a 'free time' basis, which means that it is a job for enthusiasts with a physical and mental capacity to work – if needed – three or four nights a week regularly without remuneration. Fortunately, however, ITC supported the ICA Secretary-Treasurer and President with office space and services and, on top of that, Professor Ormeling was fortunate to work with ITC secretaries of high international ability.

Only two of them may be mentioned here – both extremely valuable to ICA – Annette Kordick and Marlies Simmons.

ICA/ACI

At the age of 38, Ormeling showed up for the first time on the international scientific scene by presenting a paper to the 16th International Congress of the International Geographic Union in Lisbon in 1949, with the title, 'Progress in Map-making in Western New Guinea'. In 1958 he and C. Koeman were representatives of the Netherlands at the meeting in Mainz, FRG, to assist in the final preparation of the official foundation of the International Cartographic Association. The latter finally took place in Bern in 1959. Due to his then workload in atlas production, he did not attend the first ICA General Assembly in Paris in 1961 but, from 1962 on, Professor Ormeling attended and actively participated in every Technical Conference (13 in all) and every General Assembly (8 in all). Figure 2 gives a graphic summary of all ICA Conferences and Assemblies up to 1987.

The Imhof-Gigas founding period of the ICA produced high quality, long-term policy expressed, for example, in realistic statutes, the creation of a sound economic base for the Association and the fostering of a strong belief in cartography as an instrument for international co-operation and understanding. This positive atmosphere had a strong appeal for Ormeling. He agreed to be a candidate for the post of Secretary-Treasurer when Professor Gigas retired in 1964. Accordingly, Ormeling's official work for the ICA began in Edinburgh when he was elected to that office in the Association.

His first big task was, as Conference Director, to organise the 3rd ICA Technical Conference in Amsterdam in 1967. This was a challenge for a small country. The conference was a considerable success and, thanks to Ormeling's informal and warm welcoming address, a friendly family atmosphere in the ICA proceedings was generated and became the norm in the years to come.

The position of a Secretary-Treasurer, as a rule, involves working hard but not being seen to the same extent as the President. The hyper-active Ormeling co-operated closely with the equally active and strong personalities of Presidents Thackwell, Salichtchev and Robinson during his 12 years of office. The result of this was a flourishing Association, which continued to expand during his own Presidency between 1976 and 1984. It is not necessary here to describe further the evolution of the

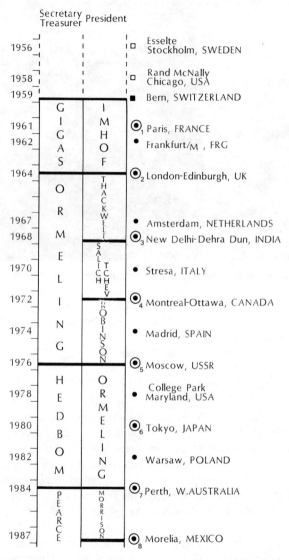

Fig. 2. ICA/ACI officers and conferences 1956–87.

Fig. 3. Fer and Rini Ormeling in his study in Lonneker, March 1983.

Association during this extended period: it is now well documented in Ormeling's book '25 Years of ICA', presented at the ICA General Assembly in Morelia in 1987.

RECOGNITION OF HIS SERVICES

As President of the Royal Netherlands Geographic Society, as well as President of the ICA, Professor Ormeling had many opportunities to present medals and awards to eminent geographers and cartographers. Upon his initiative at the Executive Committee meeting in Stockholm in 1979, the Carl Mannerfelt medal was established as the highest ICA distinction. The medal was named after the initiator of the ICA, Dr Carl Mannerfelt of Sweden. The General Assembly accepted the proposal with enthusiasm.

Ormeling himself has received many signs of appreciation, such as Honorary Member of the Dutch Cartographic Society, Honorary Fellow of the Australian Institute of Cartography, Membre d'Honneur de la Société de Géographie de France, the British Cartographic Society Medal, Honorary Fellowships of the Hungarian Academy of Sciences, the Polish Geographical Society and of the ITC. The highest distinction awarded to Professor Ormeling was given in 1978 by the Queen of the Netherlands, who appointed him 'Knight of the Order of the Netherlands Lion'. At the 8th ICA General Assembly in Morelia, Mexico, in 1987 he received the Mannerfelt Medal.

FER AND RINI ORMELING

Professor F.J. Ormeling Sr is a demanding organiser, a well-prepared, well rehearsed and fascinating speaker and entertainer, a talented linguist who speaks five languages fluently and understands several more in written form, a man with a good sense of humour and who is broad-minded. He is also the husband to a most charming and generous wife, Rini. She has supported him and assisted in solving many ICA problems and has accompanied him on almost all distant official ICA journeys. She has hosted numerous small ICA top-level meetings in their beautiful home in Lonneker, Enschede, or in their Swiss 'chalet' in Gryon. Fer Ormeling has many times been addressed as 'Mr. ICA' and Rini Ormeling has just as often been described as the 'Queen of ICA'. Together, they represent the very best of international married couples.

CONDENSED LIST OF PUBLICATIONS BY F.J. ORMELING

Progress in mapping in Western New Guinea. *Proceedings* of the 16th International Conference of the IGU, Lisbon,1949.
The Timor Problem. *Proceedings*, XVIIth IGU Congress, Washington, 1952, pp 653–657.
The Timor Problem, A geographical interpretation of an underdeveloped island. Groningen, 1955. Thesis for Doctor's degree, 250pp. Second edition, Groningen, 1957.
Wall maps for schools: The world, Europe, Africa, Australia, North and South America, Netherlands, Provinces of the Netherlands, Dutch New Guinea, Surinam, 1955–64.
De Wolters' Wereldatlas, Groningen, 1961.
Der Grosse Bertelsmann Weltatlas, Gütersloh, FRG, 1961 (Dutch edn).
Open kaart, Inaugerele Rede, Universiteit van Amsterdam, 1965, 24pp.
Attendance at Technical ICA meetings. *IGU Bulletin* 12, 1966, 41–2.
Four years ICA 1964–68. *IGU Bulletin* 20, 1969, 38–49.
Participation of ICA in IGU Regional European Conference in Budapest, 1971. *IGU Bulletin* 22, 1971, 38–9.
Report on the International Cartographic Association 1968–72. *IGU Bulletin* 23, 1972, 65–74.

Toponomy. ITC Lecture Notes for C2 students. Enschede, 1972, 25pp.

Vote of thanks to Kartographisches Institut Bertelsmann. *Intl. Yearbook of Cartography* XIII, 1973, 15.

In memory of Stéphane de Brommer. *ITC Journal* 1973–1, 175.

Turbulent cartography. Inaugural address to ITC. *ITC Journal* 1973–1, 13–37.

Introduction of Kirschbaum Verlag as publisher of the International Yearbook of Cartography. *Intl. Yearbook of Cartography* 1973, 15.

25 years ITC: a triple celebration: 17th December 1973. *ITC Journal* 1974–1, 65–70.

Cartography in ITC and ICA. *Map* (in Japanese) 12, 1974–1, 18–21.

History and significance of the Multilingual Dictionary of Technical Terms in Cartography. *ITC Journal* 1974, 705–8.

(Co-ed with A.Brown) *Geostatistics for cartographers*. ITC Lecture Notes, Enschede, 1974, 85pp.

Progress in the ICA. *ITC Journal* 1975, 161–3.

La Toponymie. *Revue de Belge de Geographie* 1974, 3–4.

ICA – a critical appraisal. *Nachrichten a.d. Kart. und Vermessungswesen, Sonderheft Festschrift Knorr*, 1975, 97–107.

A Dutch Cartographic Association. *Canadian Cartographer* 12, 1975, 222–3.

(Co-ed with E.S.Bos and C.A.van Kampen) *Proceedings* of the Seminar on Regional Planning Cartography, Enschede, 1975, 296pp.

The ICA Commissions and Working Groups and their terms of reference. *IGU Bulletin* 1976–1, 159–71.

Model for cartographic education. *ITC Journal* 1976, 367–99.

(Co-ed with R.J.M.J. Bertrand) *Atlas Cartography* ITC Lecture Notes for C2 and C3 students, Enschede, 1977, 30pp.

An appeal for solidarity. *Intl. Yearbook of Cartography* Vol XVII, 1977, 7–15.

The International Cartographic Association 1972–1976. *Intl. Yearbook of Cartography* Vol XVII, 1977, 21–31.

Some trends in modern cartography. *Nederlands Geodetisch Tijdschrift* 8, 1977, 145–8.

In Memory of Vice-President Maxim Ivanovitch Nikishov (1903–77). *Intl. Yearbook of Cartography* Vol XVIII, 1978, 13–14.

Professor K.A. Salichtchev: Honorary Fellow of the ICA. *Canadian Cartographer* 15, 1978, 100–4.

The Bos atlas, a centenarian bestseller. *Canadian Cartographer* 15, 1978, 106–13 (also in *Intl. Yearbook of Cartography* Vol XIX, 1979, 104–13).

Discours d'ouverture de la 9e conférence de l'ACI. *Bulletin du Comité Français de la Cartographie* 78, 1978, 107–10.

Welcome address to the seminar on Computer-assisted Cartography held in Nairobi, Kenya, 1978. In: *Computer-assisted Cartography*, ed. L. van Zuylen, ICA Publication, Enschede, 1979, 12–15.

In Memory of Vice-President Lech Ratajski (1921–1977). *IGU Bulletin* 20, 157–9.

Annual Report of the ITC, 1978. *ITC Journal* 1979, 1–24.

Aufgaben der Internationalen Kartographischen Vereinigung (IKV). In: *Kartographische Aspekte der Zukunft*, Bielefeld 1979, 9–30.

The purpose and use of national atlases. *Cartographica Monograph* 23, 1979, 11–23.

Cartography courses at the ITC. *Bulletin of the Society of University Cartographers* 1979–13, 37–48.

Closing address of the IXth Technical Conference of the ICA, University of Maryland, USA. *Intl. Yearbook of Cartography* Vol XIX, 1979, 19–23.

Exonyms, an obstacle to international communication. *ITC Journal* 1980–1, 162–77.

Report of the second UN Regional Cartographic Conference for the Americas, Mexico City, 1979. *ITC Journal* 1980, 364–9.

The International Cartographic Association: an information paper. *American Cartographer* 7, 1980, 5–19.

Address on the occasion of the presentation of the Mannerfelt medal to Dr Carl M:son Mannerfelt during the annual convention of the Swedish Cartographic Society. *IGU Bulletin* XXXII, 1981, 2, 53–6.

Kartographie im Wandel – Aspekte von Gegenwart und Zukunft. *Katographische Nachrichten* 31, 1980, 4, 125–39.

ICA honours Professor Sandor Radó. *Cartographica* 18, 1981, 79–83.

The President's Opening and Closing Addresses at the 10th International Conference on Cartography, Tokyo 1980. *Cartographica* 18, 1981, 68–79.

ICA Teaching Seminars in the Third World. *Cartographica* 18, 1981, 112–13.

Address on the occasion of the presentation of the Mannerfelt Medal to Professor A.H. Robinson. *Cartographica* 18, 1981, 110–12.

The United Nations and the standardisation of geographical names. In Chinese. Proceedings of the ICA Seminar in Wuhan, China. Uitgave van het Wuhan-College of geodesy, photogrammetry and cartography 1981, 1, 1–10.

Achievements in Polish cartography. *ITC Journal* 1982, 2, 8pp.

Final Project: valedictory address. Enschede, 1982, 24pp.

The Eleventh International Cartographic Conference in Warsaw, Poland, 29th July – 4th August 1982. *Intl. Yearbook of Cartography* Vol XXII, 1983, 7–18.

ICA Rendez-Vous in the Southern Hemisphere, Perth 1984. *Intl. Yearbook of Cartography* Vol XXIV, 1985, 7–22.

One hundred years of urban mapping in Rotterdam (in Dutch), Colofon Kontakten, Rotterdam, 1985.

2.Ausbildungskursus der Vereinten Nationen in Toponymie in Rabat. *Kartographische Nachrichten* 1985/3, 97–8.

ICA 1959–84. The first twenty five years of the International Cartographic Association. Enschede, 1987.

I

CONTEMPORARY REGIONAL ISSUES IN CARTOGRAPHY

THE STATUS OF CARTOGRAPHY IN NIGERIA

Isaac A. Adalemo and Olayinka Y. Balogun

HISTORICAL BACKGROUND

Before the commencement of British colonial administration in Nigeria, maps of parts of Nigeria had been drawn by European navigators, traders and explorers. Evidence also abounds to show that Nigerians too had made some contributions to the mapping of their country before British colonial rule started. Two such maps drawn by Nigerians could be preserved because they came in the reports of some British explorers. One, drawn by a Hausa man in 1797/98, was later known as Horneman's map of Hausa (Bovill, 1964). This map showed the different countries north of the Niger that later formed Northern Nigeria. The second map, with Arabic inscriptions which represented parts of Nigeria and West Africa, was drawn for Hugh Clapperton in 1825–1827 by the Sultan of Sokoto, Mohammed Bello. In the early years of British rule, there was great dependence on such maps and on those sketched by explorers, further augmented by maps compiled from boundary, mining and railway surveys.

Throughout almost the entire colonial period, most maps of Nigeria were compiled in that country but were designed for cartographic production in Great Britain by the Geographical Section of the General Staff (War Office), by W.A.K. Johnston Limited of Edinburgh, by the Ordnance Surveys and by the Directorate of Colonial (later Overseas) Surveys at various times. Only a few maps were actually compiled, designed and produced in Nigeria and this largely occurred after 1945. This background explains the slow development of cartographic manpower and culture not only in Nigeria, but also in most former British

15

dependencies. The practice of carrying out surveying and map printing in Nigeria, adopted by the Directorate of Overseas Surveys from 1945 when intensive mapping of British colonies started, encouraged the development of skilled local manpower and of equipment only in surveying and in lithographic printing. Because cartographic work continued to be carried out in Britain, no encouragement was given to the local development of cartographic manpower and equipment.

ORGANISATION OF THE MAPPING

Maps are generated through the cartographic activities of both government and private mapping agencies. More than 80 percent of maps produced in Nigeria are commissioned by government agencies. Consequently, most mapping firms registered in Nigeria are dependent on government mapping contracts.

The national mapping agencies are the Federal Surveys and the Geological Surveys. The Federal Surveys has produced and published 1:500,000, 1:250,000, 1:100,000 and 1:50,000 scale topographical maps of Nigeria, the *Atlas of the Federal Republic of Nigeria* and has recently embarked on 1:25,000 scale topographical mapping of the country. Apart from these major cartographic products, the Federal Surveys also carries out miscellaneous mapping projects for other government departments. For some time now, for instance, it has been handling the township mapping of some major Nigerian towns. The Geological Survey, on the other hand, engages in the production of specialised maps – geological maps of the entire, and parts of, the country. Maps produced by other government departments are listed in Table 1.

Most State survey departments are well equipped for cartographic work. However, they produce few maps; the reasons for this include the lack of qualified cartographic manpower, a lack of operational funds and (largely) a lack of a sense of cartographic direction, their heads (Surveyors-General) being purely surveyors and not cartographers (Balogun, 1985).

Generally, mapping activities in Nigeria require a better co-ordination so as to avoid duplication of efforts, enhance the harnessing of cartographic resources and help in the formulation of better mapping policies (Adalemo 1982, 1986). At present, several depart-

TABLE 1
Major Government Organisations Engaged in Substantial Mapping

Name of organisation	*Nature of mapping*
1. Federal Surveys	Formulation of survey policies National triangulation Topographical mapping National atlas production Aerial photography
2. State Surveys (19 states and Federal Capital Territory)[a]	Extension of triangulation cadastral survey and mapping Production of state maps
3. Geological Surveys	Evaluation of nation's mineral resources and systematic geological mapping at 1:100 000 scale
4. Federal Department of Forestry	Vegetation and land use maps of the country
5. Federal Department of Agricultural Land Resources	Soil maps
6. Nigerian Ports Authority	Nautical charts
7. Department of Civil Aviation	Aeronautical charts
8. National Population Bureau	Enumeration area maps and population maps
9. Army Mapping Centre	Military grid maps (taking over from Federal Surveys)
10. Inland Waterways Department	Maps of Nigerian rivers

[a] Since this paper was originally prepared, two more states have been created in Nigeria, bringing the total number to twenty one. However, no additional separate State Surveys have been established at the time of completion of this chapter.

ments commission similar mapping projects which overlap in geographical coverage. The establishment of a National Mapping Commission, for example, would prevent the procurement of two sets of aerial photographs at almost the same scale for the same area that is being mapped by two different departments for different (or even for similar) purposes.

THE LEVEL OF TECHNOLOGICAL DEVELOPMENT

No technology for cartographic production, with the possible exception of computer-based tools, is too sophisticated for a developing nation to adopt and apply. However, like all former colonies, Nigeria has lagged behind in this respect. All technologies required are being imported from developed countries. This is a first step towards technological development. The next step is technological assimilation and innovation which, since 'necessity is the mother of invention', is now being induced by the recent economic difficulties of our country.

As at this time, there are no firms manufacturing cartographic equipment and materials in Nigeria. Yet a full range of this equipment and material is being widely used by cartographers, architects, printers, photographers, artists and publishers and is therefore widely marketed. In contrast, the adoption of automation for the production of maps is still at the embryonic stage. It has been brought to that stage only by the interest shown by universities, even though their activities are terribly limited by lack of adequate funds for equipment and research. Even the largest official mapping agency in the country has not given a thought to the production of maps with the aid of computers. One may only hope that the computer consciousness recently being shown by the Nigerian public may extend beyond 'numeracy' and 'literacy' to 'graphicacy'.

CARTOGRAPHIC MANPOWER AND EDUCATION

The largest employer of cartographers today is the Federal Surveys Department, which is responsible for geodetic and aerial surveys and topographical mapping of the country. The second largest employer is made up of state survey departments which, as pointed out earlier, do virtually no map-making. Other employers of cartographers include the Geological Survey of Nigeria, Federal Department of Forestry, Federal Department of Agricultural Land Resources, the National Census Bureau, Geography departments of Nigerian universities and petroleum-producing companies. Mapping companies hardly employ local cartographers because they almost invariably rely on their foreign partners for cartographic design and production of their maps. Almost all Federal and State Ministries and departments rely on the Federal Surveys and state survey departments for their map supply; apparently they have

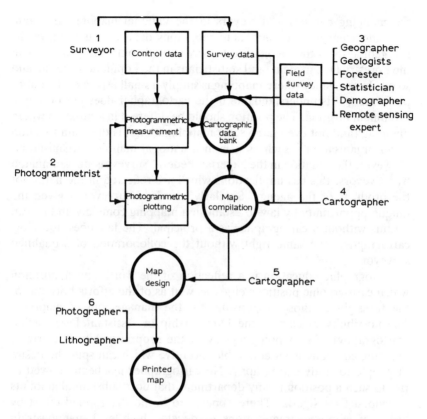

Fig. 4. The roles of professional groups in the production of maps.

little awareness of the importance of maps for the planning and execution of their own projects and the publication of their results. If it were otherwise, the Ministries would have realised that the Federal and State Survey departments can only provide general maps which should only serve as a base for the display of thematic data being collected by them.

The staff structure in the Federal Surveys of Nigeria epitomises the position of cartographers in mapping concerns in Nigeria. Several professionals are involved in the production of a map, as shown in Fig. 4. The situation in Nigeria, and apparently in many other countries, is that the professionals who provide the data (control, topographical and field survey data) for mapping tend to monopolise the management *cadre* of the mapping concern – surveyors in the national mapping and state mapping agencies, geographers in geography departments of universities, geologists in the Geological Survey and oil companies, foresters in the Forestry Department, and statisticians in the Population Bureau and so on. In an agency where mapping is simply a small aspect of the duties performed, it is understandable if the cartographer does not occupy a prominent position. The position should be different in a situation where mapping (and not even large scale mapping alone) is the main function of an organisation – such as it is in a national mapping organisation. However, the situation in the Nigerian Federal Surveys is still dominated by surveyors; this has implications which are reflected in the nature of the products of the agency. Similarly, the land surveyor is given the unique opportunity by law to establish a mapping company and he can do this without a cartographer as a principal. The law does not allow cartographers the same right without the colloboration of a qualified surveyor.

Cartographic education is a reflection of the worth and admiration which cartographic positions enjoy as well as of the efforts born out of the foresight of those responsible for the management of mapping. The possibility of gaining the Directorship or Assistant Directorship (cartography) of a mapping agency or the proprietorship of a private mapping company is an admirable incentive which can spur the desire of people to study cartography. No cartographer has been allowed to rise to such a position in any department that does substantial amounts of mapping in Nigeria. There appears to be no determined effort by heads of mapping organisations to develop high-level manpower in cartography.

At present, only one institution, the Kaduna Polytechnic, offers formal education leading to the National Diploma and Higher National Diploma in cartography. Cartographic education in Nigerian universities has, however, experienced tremendous growth recently. As shown in a previous study by Balogun (1983), between two and eight cartography courses are taught in most of the universities towards degrees in geography (see Table 2). In contrast, most surveying departments offer only one course in cartography. No degrees are awarded in cartography as of the present time. Geographers want to continue to see cartography as a specialisation in geography just like settlement or economic geography, and surveyors regard cartographers as draftsmen and, at best, technicians.

Most of the Basic Cartography Training Schools started by some

TABLE 2
Cartography Courses Offered in Geography Departments of Nigerian Universities

University	Prelim.	Part I	Part II	Part III	Postgraduate
Lagos	Practical geography	Elements of mapwork	Elements of cartography	Cartography	Cartography
Ibadan	Map reading	Map reading & interpretation	Cartography	Advanced cartography[a]	Cartography
Ilorin	Introduction to mapwork, Elementary cartography	Map analysis	Cartography	Advanced cartography[a]	Cartography[a]
Bayero	Map reading, Practical geography	Map interpretation & data presentation	Map construction & air photography	—	—
Jos	Practical geography in Africa	—	Cartography	Advanced cartography[a]	—
Calabar	Mapwork & analytical geography	Cartography	—	—	—
Benin	Practical geography	Practical geography	Cartography techniques & mapwork	—	—
Ife (now Obafemi Awolowo University)	—	Introduction to cartography, Map analysis	—	Statistical mapping, Cartography, Advanced cartography,[a] Cartography application[a]	—
Port Harcourt	—	Map reading & air photography	—	—	—
Nigeria-Nsukka	Practical geography	Practical geography	Cartography & air photo	—	—
Ahmadu Bello	—	—	Cartographic & analytic methods	Cartographic methods[a]	—
Sokoto	—	Practical geography	—	Cartography[a]	—
Maiduguri	NR	NR	NR	NR	NR

[a]Not taught due to lack of lecturers.
NR, No response.
Source: Balogun (1983).

state survey departments have died gradually because most state survey departments do not carry out any mapping. Only the Federal Surveys Department still offers regular courses leading to the Basic Cartography Certificate and Advanced Cartography Certificate. Students for these courses are drawn from the Federal Surveys, State Survey departments, universities and other government departments.

Most Nigerian cartographers, then, were trained in the USA, Canada, Great Britain, Germany and the Netherlands; a large percentage were trained in the International Institute for Aerial Surveys (ITC) in Enschede in the Netherlands. A large proportion of the people trained in Europe are now employees of official mapping agencies, while a lot of the graduates of American institutions will be found teaching in the universities and other educational institutions.

CARTOGRAPHIC OUTPUT

Cartographic production is not only a function of the need of the society but also of the training and orientation of the decision-makers in the mapping agencies. Official mapping agencies, whose management *cadre* is dominated by surveyors, tend to produce large scale basic maps such as topographical and township maps. Apart from the specialised (geological) maps produced by the Geological Surveys, there is little attention given to the production of thematic maps in official mapping organisations. The production of the *Atlas of the Federal Republic of Nigeria*, which went on for more than thirty years before it was completed, is the only major thematic mapping project carried out by the Federal Surveys.

Private mapping companies generally rely heavily on government contracts. Since governments produce mainly large scale base maps, these companies too have not helped in producing the maps which are greatly needed by the Nigerian public. A large percentage of the mapping companies carry out aerial photography and control surveys in Nigeria, a few of them do their photogrammetric plotting within the country and almost all of them carry out the cartographic production through overseas partners, thereby depriving Nigerian cartographers of much-needed employment opportunities.

LEVEL OF MAP USE

Just as few maps are actually produced in Nigeria, only a few maps are used by a few Nigerians. The latter is not necessarily a consequence of

TABLE 3
Percentage of Subjects Using Different Types of Maps

Type of map	Total number of subjects (%)
Road map	27·49
Street guide	28·82
Tourist map	9·98
Topographical map	8·87
Cadastral map	7·32
Atlas	25·50
Geological map	3·55
Soil map	3·77
Population map	8·20
Land use map	3·55
Administrative map	0·67
Historical map	0·44

the former. As a matter of fact, lack of potential demand for maps might be responsible for the lack of production of maps for the public by mapping companies – assuming that they are rational. In a map use survey, the figures in Table 3 were obtained. This shows that Nigerians, even though they use a wide variety of maps, are not intensive users of maps. Some of the maps on the list are specialised ones which would often only be used by professionals. Thus the low figures for use of geological maps, soil maps and land use maps are not surprising. But a better and widespread use of road maps, street guides and tourist maps would have been expected.

Generally, there is a close relationship between the percentage of people who use maps and the level of education. The study showed a figure of 28.9% for users whose educational background was below high school and a figure of 90% for those with post-graduate qualifications.

Institutional use of maps rates lower than individual use. The maps that will be found in those public offices which should use maps (such as the Police, Fire Services, Post and Telecommunications and Refuse Disposal) are 10 to 20 years old. Apparently, the individuals manning the offices do not attach much importance to maps. The lack of proper maintenance of public utilities has often been traced to the lack of maps showing locations of these utilities, with the result that telephone cables are often up-rooted when electric cables are being laid. In educational institutions, the use of maps has been ascribed to geographers by other students and teachers. This attitude is carried over to places of work.

Accusing fingers have been pointed by both the map makers and the map users in an attempt to explain the scarcity of maps and the relatively poor use of maps. Map makers accuse the public of being generally illiterate and not demanding maps, thus making the production of maps a risky business. Conversely, potential map users allege that appropriate and good quality maps are not available to buy: their argument is that, if the right maps were available and advertised to the right potential users, better use of maps could be effected. A case in point is the recent introduction of street guides by a mapping company. The company recorded large sales, indicating that people will use maps which are relevant to their needs.

PROFESSIONAL DEVELOPMENT

The system perpetrated by government departments of awarding mapping contracts, even for thematic maps, to non-cartographers has hampered professional development in cartography. The situation is reminiscent of the conversation between a public officer and a general contractor who was looking for contracts.

> *Contractor*: Officer, I will be happy today if you can give me any contract – at least to keep me going.
> *Officer*: Sorry, there is no contract you can handle now.We only have a map to produce and you cannot do it.
> *Contractor*: A map? What is a map?
> *Officer*: It is like the one hung on the wall over there (pointing to a wall map).
> *Contractor*: So, this is a map! Yes, we can handle it. We have an overseas agent who handles jobs like that for us.

Cartographers have competed with and lost much work to such general contractors.

There is however another problem; jobs are lost to overseas concerns even though there are competent local professional cartographers. Unfortunately, printers who can demonstrate high quality map printing are few in number. To a layman, the cartographic and the printing aspects of map production are inseparable. Sometimes, therefore, the need for flashy, beautifully printed maps has robbed cartographic professionals of opportunities to take part in the production of such maps since contractors would prefer to send the entire production of such maps to places where good printing could be done.

Most of the problems related to professional cartography are well recognised and are discussed at the meetings of the Nigerian Cartographic Association, which was founded in February 1978. The introduction of the *Nigerian Cartographer*, the official journal of the Association, will further enhance its efforts to enlighten the public about cartography and the making of maps. It will also contribute towards the achievement of its aims and objectives: these include the advancement of the art, science and technology of cartography, contributing to public education in the use of maps and other cartographic products, encouraging improvement in cartographic curriculae of universities and polytechnics, supporting programmes of cartographic research, seminars, workshops and publication and fostering good relations between members of the Association and other surveying and mapping and geographical associations.

The recent economic hardship in Nigeria has made it difficult to pay for cartographic services handled abroad; thus more jobs are now open to cartographers who are themselves becoming more enterprising and more aggressive than ever before. Several cartographic services have been set up in the last three years and it is hoped that they will be able to prove their competence before the Nigerian currency appreciates in relation to other currencies and again makes it easy for services to be purchased from overseas which could be provided locally.

CONCLUSIONS

Cartography, an old profession, is paradoxically a slow starter as an academic discipline. Nigeria is one of the many countries where academic qualifications matter a great deal in securing jobs and in gaining professional recognition. Most academic programmes leading to diplomas/ degrees in cartography all over the world were only initiated within the last twenty years. They are yet to start in Nigeria. This situation, which can be traced to the manner in which Nigerian maps were produced in the colonial days, has resulted in the unenviable position in which the Nigerian cartographer still remains. The situation will change as Nigerian society becomes more literate, its environment becomes more complex and more maps are needed to sort things out and create 'order on the land' (Mabogunje, 1980).

At present, the official mapping agencies still follow the mapping processes and systems left for them by their former colonial masters without any notable changes. The title of 'survey draftsman' has merely been changed to 'cartographer'. As Keates (1985, p.27) also noticed with respect to Great Britain, the official mapping agencies regard these cartographers as technicians 'with specialised drafting skill'. However, once the stereotyped mapping procedure is changed, the need for scientific and professional cartographers will be realised. In most advanced countries, maps – especially the official ones – are being re-designed to satisfy the requirements of the map user. Research is being conducted into the use of the computer for rapid production of maps and in the use of new data sources. All these developments are bound to demand the services of highly qualified cartographers. Innovations will be brought into the field of mapping. Better and larger varieties of maps will be produced and a greater number of people will use maps.

REFERENCES

Adalemo, I.A. (1982), *Coordinating Mapping Activities in Nigeria*. Presidential Address delivered to the 4th Annual Conference of the Nigerian Cartographic Association, University of Lagos.

Adalemo, I.A. (1986), *Harnessing Mapping Resources for Development in Nigeria*. 1986 Olumide Memorial Lecture, Nigerian Institution of Surveyors, ASCON Topo – Badagry, September.

Balogun, O.Y. (1983), 'Structure and Problems of Cartographic Education in Nigeria'. *International Yearbook of Cartography*, 23, 121–35.

Balogun, O.Y. (1985), 'Directing the State Mapping Programme: Surveyor-General or State Cartographer'. *Proceedings of the 7th Annual Conference of the Nigerian Cartographic Association*, October 9–11, Kaduna, 1–13.

Bovill, E.W. (Ed.) (1964), *Missions to the Niger, Vol. 1. The Journal of Friedrich Honemann's Travels from Cairo to Murzuk in the Years 1797–98 and the Letters of Major Alexander Gordon Lain 1824–26*, Cambridge.

Keates, J.S. (1985), 'Cartographic Education in the Mapping Science Field', in D.R.F. Taylor (Ed.), *Education and Training in Contemporary Cartography*, John Wiley & Sons, 27–52.

Mabogunje, A.L. (1980), *Order on the Land*. Keynote Address to the 2nd Annual Conference of the Nigerian Cartographic Association, June 26–28, University of Lagos.

THE STATE OF CARTOGRAPHY – THE INDIAN PERSPECTIVE

G. C. Agarwal

INTRODUCTION AND BRIEF HISTORY

Descriptive mapping is a very old activity in India. More than 5000 years ago, maps were present in the Vedic literature – not in graphical form but by description of the extent and shape of territory. The basic concepts of modern map making, such as scale and generalisation of features, were however known in India at an early date, as is evident from ancient Puranic literature of 500 BC to 700 AD (for example in the *Brahmananda Purana*). It has been established that the art of surveying and techniques of mensuration of areas were well developed in ancient India; the manual known as the *Sulva Sutra* (the Science of Mensuration) bears testimony to this early knowledge. References also exist in this regard in the Third Century BC *Artha Sastra Chanakya*. In the medieval period, Sher Shah Suri's revenue maps, based on a regular land survey system, were well known; this system was in vogue until the mid-eighteenth century.

In 1767, the East India Company established a survey department in Calcutta, in the Bengal Presidency, to survey their acquired possessions. Later, as the influence of the company grew in other parts of India, similar survey departments were established in the Bombay and Madras Presidencies in 1796 and 1810 respectively. Subsequently, the three departments were merged in 1815 to form one single department and named 'the Survey of India'. Among the notable achievements of this Department during its initial years was the Great Trigonometrical Survey. This provided the physical constants pertaining to the shape of the earth and the basic geodetic framework for all survey and mapping activities in the country. The Department also covered, by topographical surveys on scales of 1 inch to 1 mile or smaller, the accessible parts of the country. Until the end of the nineteenth century, maps were produced in one colour – black – with topographical features shown by hachures. Cartography grew as an adjunct of this mapping.

In 1905, a high powered committee set up by the Government of India recommended that the primary map scale of 1 inch to a mile (1:63,360) be adopted for the whole country, with depiction of topographical features by contours, and that the maps be printed in multiple colours. The revision cycle was fixed at 25 years. The Revenue (cadastral) surveys and mapping were uncoupled from the Survey of India and became the responsibility of the provinces (now states) on the recommendation of the same committee.

From that date until Independence in 1947, surveying and mapping activities were carried out under the direct control of the British on the primary map scale on 1 inch to 1 mile, but difficult and inaccessible areas, including the frontier and trans-frontier areas, were mapped/compiled on a scale of 1/2 inch to 1 mile (1:126,720) or 1/4 inch to 1 mile (1:253,440). Major set-backs to this activity occurred during the World Wars as all survey potential was diverted to the war effort. By 1947, only about 60% of the country was covered by maps at the primary scale of 1 inch to 1 mile.

A significant role which the Survey of India was called upon to assume emerged soon after the country attained Independence. With the adoption of a planned economy as a national objective, a phenomenal demand emerged for accurate positional information about natural and cultural details required to initiate multi-purpose developmental projects. Since the Survey of India was the only organisation equipped to undertake such massive tasks, the responsibility was assigned to it. The promulgation of the Metric Weights and Measures Act of 1956 added yet another dimension to the work of the Survey in requiring a change-over from the FPS or Imperial system of mapping to the metric system. Thus the scales on which maps were to be surveyed and printed became metric, namely 1:25,000, 1:50,000, 1:250,000, etc. in place of 2 inches to 1 mile, 1 inch to 1 mile and 1/4 inch to 1 mile, and so on.

With increasing sophistication in the fields of aviation, communication, meteorology, hydrology, forestry, tourism, urban and rural development, environmental planning and in education, the demand for cartographic products multiplied over the years. Education and training in cartography, which were mostly confined to be within the Survey of India, were the main vehicles for the growth in numbers of skilled cartographers in India. This growth has been augmented by making the training facility available to others in India and abroad, following the establishment within the Survey of the Centre for Survey Training and Map Production in 1965 at Hyderabad with UNDP assistance. This centre has been instrumental in bringing in the latest technology in processing of geographical data from aerial photographs, replacing the conventional pen and ink drafting by scribing and introducing modern reprographic processes.

The other government departments which have developed special cartographic products to meet their own requirements are the Geological Survey of India and the Naval Hydrographic Department. Amongst the relatively new agencies which have appeared on the national cartographic scene are the National Atlas and Thematic Mapping Organisation (NATMO) which has produced a very good National Atlas Series at scales 1:6M; 1:2M and 1:1M, the National Remote Sensing Agency (NRSA), the Census Map Office (CMO) and a few universities and commercial agencies. A number of regional centres for resources mapping have been set up. The Centre for Earth Science Studies (CESS) in Kerala (a southern state of India) is one such centre. Thus regional effort in developing cartographic potential has started. A number of planning atlases at the regional level have also been brought out and the trend is likely to continue.

There have been significant developments in establishing training courses for the ever-increasing demand for well-trained cartographers. Postgraduate level courses are now available and a few universities have established Chairs of Cartography. Thus scientific growth of the discipline is assured in India.

MAPPING METHODOLOGY IN INDIA

In India, 'surveying and mapping' means the whole range of operations connected with determining relative positions of points above, on, or beneath the surface of the Earth, and – after establishing such points – collecting related terrain information and depicting the same on paper. In a general sense, surveying has been regarded as that discipline which encompasses all techniques for gathering and processing information about the physical earth and the environment.

The conventional surveying and mapping activity comprises the following four distinct operations:

(i) Ground control work
(ii) Survey of specific information
(iii) Cartography
(iv) Reprography

Control Work

The surveying process is normally initiated by establishing a framework of points of a high degree of accuracy, in all three dimensions, in order to maintain the graphic accuracy of features on a national scale and – if possible – on a global scale.

The basic framework for India was established through Geodetic Triangulation by measuring angles and lengths of spatial geometric figures formed by discrete observation stations at prominent terrain locations. This consisted of chains of basic triangles with an average side length of 90 to 100 km, set up in the form of a grid. The distance between any two adjoining chains was about 400 km. Vertical control was provided by High Precision Spirit Levelling. Along with the establishment of this basic framework, the study of the physical parameters of the earth (such as size and shape and the strength of the gravity field) was also carried out using direct measurement techniques such as triangulation, precise traversing, levelling and gravimetric observations by High Precision Spirit Levelling.

The 'work horses' of geodesy in the past were the huge and heavy theodolites and other equipment which had to be carried in pieces and assembled at the site of observation. But, with the advance of instrument technology and optics, the theodolites have been designed to be easily portable and capable of giving even higher accuracy. The first order geodetic control provided by use of these techniques needed to be 'infilled' through second and third order triangulation, traversing and levelling so as to provide adequate and independent control for survey of the area of each map sheet. Thus positional errors in an area are controlled and localised to remain within permissible limits by the peripheral, primary geodetic control points.

Survey

Establishment of control points is the start of the survey process. The subsequent steps are:

(i) To survey and verify ground details,
(ii) To annotate and introduce names of rivers, roads, villages, towns, etc. (also called semantic information),
(iii) To depict topographic features by contours and other suitable symbols.

Survey operations in the past were carried out through the ground technique of plane-tabling, in which details are fixed by resection and intersection of rays, and sometimes by direct chaining, with contours interpolated from height information. This process was laborious and time-consuming. With the emergence of aerial photography, the conventional surveying techniques underwent a dramatic change. The Survey introduced these new techniques and thus reduced substantially the rigours of ground survey. Almost all surveys by the Survey of India were carried out using the so-called air-cum-ground method, in which the basic data obtained from aerial photographs by graphical methods were supplemented by ground truth collected from field visits. This method also resulted in considerable reductions in the requirement for ground control.

By the early fifties, the graphical method of survey from air photographs was replaced by the use of precision analogue photogrammetric instruments. The first photogrammetrical machine, a Wild A7, was installed in India in 1954 with UN Technical Assistance. This was followed by the acquisition of a host of analogue restitution instruments of different orders. Instruments for rectification and orthophoto production for special mapping projects were also procured. Currently, analytical stereo-plotting instruments are being introduced in the Survey of India.

Generally, the aerial photographs used in the Survey as a cartographic data source are vertical or near-vertical. However, use of high oblique photography for data acquisition of inaccessible areas was successfully used during the Second World War for small scale mapping. Since that time, vertical aerial photographs have become an increasingly important tool in the hands of a surveyor. But, although analogue photogrammetry had reduced ground work considerably, field visits were still inescapable. This complementary ground survey took two forms:

(i) Minimal densification of existing ground control and its identification for photogrammetric triangulation to whatever extent was necessary.
(ii) Ground verification of information not obtainable from photographs, such as detail in 'dead ground', under cloud or vegetation cover, the names of places, hills and rivers and the location of post offices. Such a verification was carried out either by taking the photographs to the ground and

annotating them or taking the results of a photogrammetric survey to the ground for supplementing and editing.

In the Survey of India, the process of map-making with aerial photographs begins with planning of the aerial photography. The ground control points at selected locations are connected with existing geodetic control by field survey and their geographic positions are computed. The 'signalisation' of ground control points is done so that their identification on photo imagery is easier. If this has not been possible before the photo-mission, then photographs are taken to the field at the ground control stage for marking up the control on the photo imagery. With the photographs and with the minimal ground control and their image positions identified, the stage is set for undertaking independent model photogrammetric triangulation and subsequent computations and adjustments on digital computers. This provides the requisite ground-related control points in each model. The models are properly oriented and then used for generating cartographic data. At a convenient stage, collection of ground truth and supplementary ground information are undertaken. The verified data are used to edit the cartographic data obtained from the models.

Cartography
In the chain of steps in map-making, cartographic activity pertains to the editing of survey data (in the form of details of cultural and natural features and contours) collected through photogrammetric or ground survey methods. It also includes the preparation of master originals for the final stages of map reproduction. In the conventional method, the survey sections are photographed and the black prints obtained are mosaiced for the area covered by each map sheet (a $15' \times 15'$ quadrangle for 1:50,000 scale sheet). Light-blue prints of this mosaic are obtained on drawing paper or on scribe-coated mylar sheets, from which colour-separated pen and ink drawings or scribed negatives are made. These are subsequently used for preparing printing plates.

Smaller scale maps are compiled from larger scale ones. Thus, to obtain a 1:250,000 scale map from 1:50,000 scale maps, sixteen of the latter are individually edited manually and details generalised before obtaining photographic reductions at 1:250,000 scale. Black prints of such reductions are mosaiced to cover the area of the 1:250,000 scale sheet and fresh prints are obtained on drawing paper or scribe-coated mylar sheets for preparing the colour. This process is followed to build up the entire family of map series down to 1:4M and smaller scales.

Printing
The map manuscripts are colour-separated where necessary and press-plates are prepared through a chain of intricate reprographic processes. The final printing is carried out on rotary off-set multicolour printing machines.

PRESENT TRENDS

Revolutionary developments are taking place in survey technology based on developments in the various related fields, particularly in electronics, computer technology, space science, remote sensing, graphics and printing. Because of this, cartographic products such as maps and geodata – being means of communication of geo-information – have to undergo changes in the way they are produced. Keeping this in view, the technology available in the Survey of India is being continuously up-dated to meet the requirements of development surveys and topographical surveys.

Distance is now generally measured with precise electronic ranging instruments. Angles continue to be measured with a theodolite. The introduction of Electronic Distance Measuring Instruments in the 1960s has simplified and expedited the work of establishing ground control for photogrammetric block adjustment and for large scale surveys. Using the conventional terrestrial system, a major constraint was the limitation to line-of-sight measurements which could only cover a range of 80–100 km. However, with the advent of an elevated platform given by satellite technology, it is now possible to relate points 1,000 km apart. The system in use at present in the Survey of India is the US Navy Navigational Satellite System. In India, satellite geodesy is mainly used for three main purposes. These are the strengthening of the existing network, densification of control (including provision of control in inaccessible areas) and linking in of the off-shore islands.

In the conventional analogue photogrammetry followed by the Survey of India, computers have been helpful in computing and adjusting ground control and in carrying out photogrammetric triangulation in a semi-analytical mode. The present trend in the Survey is to automate various stages of photogrammetric survey and thus expedite the cartographic output. In view of this, analytical plotters such as the PLANICOMP C–100 are being introduced. With modern analytical plotters which work under computer control, not only is photogrammetric triangulation carried out rapidly but also subsequent plotting (though done by an operator) is being carried out quickly and precisely. Photo-grammetric data may thus be captured either from analogue plotters or in digital form from analytical plotters. In the latter, the operator plots a model after completing orientations and simultaneously structured digitised data are generated for subsequent editing and storage. The Survey has realised the advantages of such digital systems to meet the demands for supply of data or maps on a 'turn-key' basis and the flexibility available in the choice of the final output format; because of this, a research initiative has been set up into Digital Terrain Models (DTM) for a number of nationally important and interesting projects. These include the selection of rail route alignments, digital rectification of air photos and automated plotting (planimetry) and the preparation of slope maps. These DTMs were originally built up on the Hewlett Packard micro-processor, interfaced with digitisers and a Calcomp Flat-bed Plotter at Hyderabad, and were then transferred to the sophisticated 'Automap' system installed at Dehra Dun.

Remote sensing instruments, such as multi-spectral scanners on space platforms, provide unstructured data either in the form of digital tapes or images. Remotely sensed data may also be acquired from infra-red (black and white or colour) cameras, thermal infra-red scanners, multi-band cameras, multi-spectral scanners, microwave radiometers and radar carried on aircraft. These data can be read, enhanced, geometrically corrected and then stored unstructured in the data base. As imagery in digital form with higher resolution (like that from the SPOT satellite, for example) becomes available, digital remotely sensed mapping will become quite helpful for selected themes in small scale topographical mapping. The Survey of India is keenly interested in these possibilities and is experimenting with SPOT and other comparable imagery for topographical mapping.

Side by side with the up-dating of the survey technology, 'indi-genisation' of surveying, drawing and printing equipment has been a characteristic of the last two decades. A modest beginning had already been made by manufacturing substitutes for imports. As of now, drawing instruments and material, such as scribe-coat sheets, peel-coat sheets, drafting films and scribing instruments of good standard, are being manufactured in the country and are in regular use. On-going research and development activities are continuously improving their quality. Theodolites, levels and other survey equipment are already being design-ed and fabricated by various Indian manufacturers. A long-range EDM instrument, designed jointly by the Survey of India and the Electronic Corporation of India Limited, is in use. Mini-computers and microprocessors for processing survey data are now available in the Indian market.

AUTOMATION IN CARTOGRAPHY

With the advent of the computer and its very rapid development, Computer-Assisted Cartography (CAC) has been found to be the most important technical innovation in the recent development of cartography. It is being widely advocated as the inevitable form of map-making in the very near future, in which selected phases of the production processes or even the whole process will be controlled by the computer. In a bid to accept and adopt this technique and to modernise the mapping methods in India, the Survey of India acquired a CAC system known as 'AUTOMAP' in 1981. In this system, geographic and other data are stored in digital form and processed on demand to draw the required map output on any scale by using a computer-operated drafting table. The system meets many of the desired functions in a modern mapping organisation such as data storage, up-dating, retrieval and data translocation and yields a wide range of cartographic output, hitherto impracticable through manual procedures. Important projects which have already been carried out using this system are:

(i) Processing of complete 1:25,000 scale topographic maps for printing in colour
(ii) Computer programs for plotting map borders and graticules
(iii) Preparation of a 1:8M scale base map of India for thematic mapping, etc.
(iv) Preparation of a National Digital Data Base on 1:4M scale
(v) Processing of maps on 1:50,000, 1:250,000 and on 1:500,000 scales
(vi) Generation of DTMs and application of them
(vii) Experiments with Geographical Information Systems/Land Information Systems
(viii) 3-D perspective analysis using computer graphics

CARTOGRAPHIC PROCESSES AND PRINTING

Conventional pen and ink drafting methods of producing originals are quite cumbersome and the originals produced are not very durable. The modern methods of negative scribing and photo-lettering are much more convenient and economical in the long run. Proofs are also prepared easily by the 'Rub-on' method for editing. These methods have already been adopted for bulk and regular mapping activities. This has brought about a reduction in the total number of stages, particularly in photography, colour separation and in proving. In addition, it has given the maps a 'better look', with superior colour registration.

Apart from the introduction of scribing, the unique method called 'Colour Trol' has made it possible to produce thematic maps with numerous variations in colour shades to represent different symbols on the map. This method uses only four master plates, one each for magenta, cyan, yellow and black, and percentage screens (ranging from 30% to 100%) dropped into open window masks for superimposition of colour dots of varying densities. This is a very economical method for producing multi-colour maps of high quality and has been tried successfully in the Survey of India for production of atlases and for other special mapping projects carried out in collaboration with other government agencies and national universities.

The symbols used in topographical maps published by the Survey of India have long been standardised and served the map readers well. But, in view of the increasing and changing map clientele, a need has been felt to modify some of the symbols and the semantic content of the map. The Map Review Board, constituted by the Government of India in 1977, has recommended changes in symbols and also in the layout of the standard topographical maps on 1:25,000 and 1:50,000 scales. New editions of the sheets in the revised layout with new symbolism are already in the course of production. Exercises to devise new sand feature symbols suitable for automation and printing of them by the photo-litho process have also been carried out.

The development of a plastic relief maps is a new complement to the normal map. Until recently, production of maps in the country was confined to those of only two dimensions. In 1974, the Survey of India set up a research team in the field of three dimensional mapping. This research team consisted of a few technicians; working from first principles and using completely indigenous materials and techniques, it designed and fabricated its machinery and has since produced a number of plastic relief maps.

In addition, the Survey is deeply involved in the preparation of small scale geographical maps, atlases and user-oriented products produced with a fast response. These are designed to meet social and national needs. One example produced by the Survey is the *School Atlas*; this is very popular because of its rich content and low price (less than one American dollar). The production of a regional economic atlas about a decade ago has also set a trend and many regional atlases have since been produced and others are in the course of production.

A PERSPECTIVE ON THE FUTURE

With the completion of the 1:50,000 scale topographical survey of the whole country in 1982 and mapping from it in 1985, India has joined the select group of countries which have a complete map coverage on a national scale. It should be mentioned that, apart from the European countries which have relatively small national territories, India is perhaps the only large developing country which has established such a coverage. The Survey of India is really proud of this achievement. Considering the extent and the rich variety of information available in the sub-continent which has to be presented in graphical form, this is a monumental work (including the 1:250,000 scale compilation) which has taken 25 years to complete. Included in this is also the revision of developed areas and bringing their map coverage up-to-date by producing newer editions.

It has now been decided to revise the 1:50,000 scale series over a cycle of 5 to 10 years, depending upon the changes taking place in the information content of a map. The Department is also now embarking upon an ambitious programme to cover the whole of India by maps on a 1:25,000 scale by the middle of the 1990s. This is possible in view of the revolutionary changes taking place in the field of cartography. Techniques, technology and philosophy are undergoing kaleidoscopic changes, even though cartographic principles have remained relatively stable. Many of the most imaginative and innovative developments are driven by commercial, military and political pressures. Indian cartographers are fully convinced of the potential of these changes for improving the management of utilities, the environment, land use, social planning and the like.

In India, cartography is not a refuge for stray academicians. There is considerable excitement among Indian cartographers, who have in the past mostly played supporting roles to the dominating field workers, over the sophisticated changes which their discipline is assimilating. Gone are the days of the conventional pen and paper cartography. In the future, they will work in a dust-free air-conditioned chamber, using powerful mainframe or micro computers and peripherals and terminals with clinical precision. The future of these staff can no more be considered within a narrow traditional definition and bounds since their discipline will be interfaced with Survey, Photogrammetry, Remote Sensing, Electronics, Telecommunications and Reprography. Moreover, cartography – while revitalising itself – will contribute much to these other disciplines. In India, for instance, we are planning the birth of a new generation of cartographic practitioners in an electronic-cum-information technology environment. They will be an amalgam of surveyor, cartographer, photogrammetrist, remote sensing specialist, reprographic technologist and computer software and system expert. They must be capable of producing instant data and information packages, precision-corrected in real time in a variety of user-oriented bases and formats. As a recognition of this trend, the International Union for Survey and Mapping (IUSM) emerged as a concept in 1985 for providing interaction and coordination among the scientists of ICA, FIG and ISPRS within the scientific community of the world.

Geographic Information Systems (GIS)
We are witnessing an age where people need information of all types and related to a geographical location in a 'fast response mode': land, its uses and topography, houses, population, weather and much else. Such queries of a multifarious nature cannot be satisfactorily handled except through previously collected and stored information which can be accessed in real time. It has now become possible to do this using high speed computers through Geographical Information Systems or GIS. Basically, a GIS is a computer hardware and software system designed to collect, manage, analyse and display spatially referenced data. We as cartographers will, however, be primarily concerned with land use and topographic information. Tasks which will be involved include gridding, contouring, feature extraction, overlay, linking locations to attribute files and up-dating already stored information as new observations are made. This part of GIS could more appropriately be called a Land Information System (LIS). Cartographers will have to form the nucleus of these information systems.

GIS in India
In India, action was initiated in 1981 by the Department of Science and Technology (DST) to create a GIS. Designed to meet the information needs of many users, this involved various departments such as the

Survey of India, the Indian Space Research Organisation, the Space Applications Centre, the National Informatics Centre, the Department of Electronics. The program subsequently got a great boost through the liberalisation of Government policy on computers. In the Survey of India, we have passed through the 'stage of initial apprehension' and of replicating existing products through CAC techniques on a pilot study basis. We are now in the stage of implementing a national GIS/LIS through our newly established Modern Cartographic Centre (MCC) and Digital Mapping Centre (DMC).

It may, however, not be possible to switch completely to CAC methods on account of various constraints, particularly of resources. For us, computer-based cartography is highly capital-intensive. Both classical cartography and CAC will have to co-exist, the former giving way to the latter in due course. There are over 5000 sheets in the 1:50,000 scale map series covering the whole of India and the 1:25,000 scale is becoming our national primary scale; though it is an immense task, we should be able to create an LIS hooked on to the national GIS in the next two decades. Such prediction is difficult because CAC is in a stage of rapid evolution. It has already progressed so considerably during the last two decades, that it is impossible to anticipate its development over the equivalent time period. Perhaps Intelligent Knowledge-Based GIS (IKBGIS) will be commonplace at that time.

For the time being, we in India are planning to have a data bank on a number of scales – 1:4M, 1:1M, 1:250,000, 1:50,000 and 1:25,000. The larger scale data (1:50,000 and 1:25,000) will, when generalised, be consistent with smaller scale data files within the resolution of those small scale data. Linear and point details, such as boundaries and trigonometric points, will have unique coordinates for all scale data. That is, there will be some data which will be 'scale-free' and other data will be scale-related but consistent. Of course, as in any pioneering scheme, provision will be made to permit modifications based on later experience!

CONCLUSIONS

From the present state-of-the-art of cartography in India, it can be stated with satisfaction that the future of the discipline in our country is very promising. It is hoped that activities in private cartographic agencies, educational institutions and renowned professional bodies, such as the

Indian National Cartographic Association (founded in 1979), will grow. Greater interaction within the cartographic community, through professional gatherings and publication of literature, is likely to occur. All this bodes well for the growth of cartography in India.

Like others, the Survey of India is on the threshold of the CAC era yet has to tread with judicious caution. We are not switching over to CAC with haste. We are planning to have our LIS based ultimately on 1:25,000 scale surveys, with 'spin-off' versions of it having a reduced data content and including important features only where necessary. This would be developed as a small-scale data bank and would be the base for an Intelligent Knowledge Based Geographic Information System. We are planning to be leading partners in the development of information technology in India.

THE STATE OF CARTOGRAPHY: POLISH PERSPECTIVES

Marek Baranowski

Cartography in Poland can best be described by considering a few fields of cartograpsheric activities. First of all, the state of research work will be characterised. Then the most important achievements in practical cartography will be presented. The system of cartographic education and the Polish contribution to international cooperation will be discussed. Finally, the perspectives facing cartography in Poland are defined. It is, of course, quite a difficult task to present in this short chapter all the relevant topics, so only the most important cartographic activities pertaining to the state and to the future of Polish cartography will be considered.

RESEARCH WORK

In the last 20 years, the most dynamic trend in cartographic research in Poland has been the theory of cartographic communication. The most outstanding representative of that subject, Professor L. Ratajski of Warsaw University, developed the idea of a cartographic form of information transmission. He formulated a concept of 'cartology' as a system of theoretical cartography, viewed from the standpoint of the function of cartographical transmission of information and regarded as a theoretical superstructure of applied cartography. Thus,

> '. . cartology comprises research on the transmission process of information concerning the spatial distribution of facts, using cartographical products, the detection of rules, the determination of principles and methods for

41

optimisation of the cartographical transmission, as well as the place of cartography within the remaining scientific branches.' (Ratajski, 1970)

Professor Ratajski was also involved in an investigation on map language grammar, which he used as the basis for establishing standards of cartographic signs. He described map language as an ideographic language, capable of transmitting chorological information by use of a system of active signs (Ratajski, 1971, 1973, 1976). His work had a great impact on the development of theoretical research in cartography.

In recent work, we can find some continuation of Ratajski's ideas but research in that field is not so intensive as formerly. The problem with the place of cartographic transmission in the circulation of general information was considered by W. Grygorenko (1982). The same author has discussed the role of cartography, based on Salichtchev's concepts, and has presented the evolution of the sense of a map from being merely a description of surveys towards a source and research document for comprehensive interpretation (Grygorenko, 1984). Ratajski's research on map language was also continued in some aspects by Grygorenko (1973, 1974, 1982) and by J. Golaski (1982). Grygorenko analysed the consistency of the graphical structure of cartographic image in relation to the structure of the reality being mapped. Golaski studied the problem of names on maps from the viewpoint of sign theory.

We can observe in Polish cartographic research some influence of Salichtchev and Berliant's work on so-called 'cartographic methods of investigations'. That trend in cartographic theory is based on searching for relationships and structure in spatially varying phenomena by analysing maps. Trafas *et al.* have differentiated two research directions in the subject, i.e. to elaborate such methods of cartographic analysis and presentation which make it possible to find relations between phenomena, and to analyse relations between phenomena on the base of graphic image features.

One of the most broadly developed currents of cartographic research work is cartographic methodology. In the last decade, much research has been strongly influenced by the theory of cartographic transmission, the theory of signs and experiments on map perception. Methods of cartographic presentation were classified and thoroughly described by Ratajski (1963, 1973). He developed the concept of a so-called 'generalisation node' (1967, 1973), which was defined as a 'moment of methods' conversion, caused by change of scale. The most popular method among researchers has been an isarithmic one. The problems of statistical data display, especially using isolines, has often been analysed

(Moscibroda 1975, 1978, 1982; Ratajski 1975; Szewczuk 1978, 1983). In addition, the idea of establishing geometric standard areal units for Poland has been discussed repeatedly. The last proposal (Podlacha, 1983) was based on use of a geographic grid.

A second method of great interest to Polish cartographers has been the choropleth one. Multiple studies on the construction and characteristics of the method were undertaken by Paslawski (1980, 1982). He analysed class interval settings and worked out his own proposal, taking into account the difference of values between neighbouring statistical units.

At Warsaw University, theoretical research and experimental work was carried out on the subject of cartographic generalisation. Baranowski defined mathematical rules for the generalisation of settlements and described a selection algorithm (Baranowski and Grygorenko, 1974). Libura considered a problem of automated shore-line generalisation (Grygorenko and Libura, 1977).

A number of technology-oriented research studies have also been undertaken. At the Geodetic and Cartographic Data Processing Centre (GCDPC), an experimental thematic mapping system was developed. An administrative map of Poland at a scale of 1:500,000 was digitised in order to provide data for a cartographic data base comprising administrative boundaries, settlements, rivers and streams, lakes and shorelines. The data base had a network structure. Data were verified by means of a digital plotter (for accuracy checking) and of a graphic display (for completeness and feature code checking). Originally developed at GCDPC, an interactive subsystem was used for graphic data editing. Software development made possible processing of spatially-oriented statistical-type data in order to present them in the form of choropleth maps, diagrams or symbols. Source numerical data were ordered and classified by several procedures, all with the possibility of user intervention. Then graphic patterns and symbols were constructed by means of computer procedures. Finally, a digital plotter was used for graphic image drawing. The system was used experimentally for preparing a set of thematic maps for functional analysis of settlements (Baranowski, 1980).

A system for thematic mapping was developed at the Irrigation Design Bureau and Institute of Geodesy and Cartography. Agricultural, soil and elevation data were collected in a grid unit of the size 250 m by 250 m. Results of computer-assisted analyses were presented in graphical form by means of a lineprinter or a drum plotter. Generalisation procedures were developed, so the output maps could be produced at smaller scales than the original material used to produce the contents of the data base (Ostrowski and Podlacha, 1980).

The next computer-based technology was also developed at GCDPC. In 1984, a system to create a digital elevation model (DEM) was developed. It was assumed that the model would consist of terrain elevation values at regularly distributed points. These values in digital form make up the basic file of an appropriately organised data base. In the data acquisition stage, contour lines from the 1:10,000 scale topographical maps are digitised and preliminary data processing is performed in order to create an intermediate data base containing a digital record of these contour lines. Interactive software was developed for data base editing. The next step consists of transformation of the contour lines into a network of regularly spaced points of elevation values. Intersections of contour lines and profile lines are computed; points of intersection of contour lines and profile lines, which converge with the above mentioned regular net, are first identified. These points are then used to determine terrain elevation at the adjacent nodes of the regular network. Verification is made just after preliminary processing as well as after the interpolation of the node values. This process can be performed entirely by digital means and, by producing new contour lines, provides a check through comparison with those on the original map (Baranowski, 1984). The elevation model is created mainly for telecommunication purposes but its application will, of course, be substantially broader.

A few mapping facilities have been developed at the Remote Sensing Center of the Institute of Geodesy and Cartography. Data from air photos and satellite images were interpreted and, according to the theme, were displayed in preliminary form (as a photographic copy). The mapping process was then done traditionally. One of the results of such work is a photographic map of Poland at a scale of 1:750,000 based on Landsat images. Newer technology (in the process of preparation) will allow map making from the air by means of computer (Podlacha, 1986).

CARTOGRAPHIC PRACTICE

In Poland, the Head Office of Geodesy and Cartography is responsible for the state and development of cartographic activities. Additionally, apart from that organisation, there are many institutions which produce thematic maps in a more or less coordinated way. The Head Office of

Geodesy and Cartography, through subordinate enterprises, provides users with base maps and topographical maps of the country. The Head Office is also involved in the preparation of some thematic maps and atlases.

The national base map is being created at scales of 1:500, 1:1,000, 1:2,000 and 1:5,000, depending on the density of detail in the terrain. At present, the map exists for the greater part of the country (over 80% of the urban areas). The largest scale of topographic maps is 1:10,000, except for a few regions of special industrial importance for which maps at a scale of 1:5,000 exist. The country is totally covered at the 1:10,000 scale and the revision cycle of this map is 10 years. In addition, there are many editions of 1:25,000, 1:50,000 and 1:100,000 scale topographic maps. Recently, a map at a scale of 1:200,000 was published.

Some enterprises subordinated to the Head Office of Geodesy and Cartography have taken part in the development of a 1:50,000 hydrographic map of Poland. The mapping processes are carried out under the supervision of hydrologists from scientific centres. The work is expected to be complete within about 15 years. The Geological Institute is conducting work on a detailed geological map at a scale of 1:50,000. It is one of many cartographic undertakings of the Institute and has already been carried out for half a century. As far as soil mapping is concerned, a survey at a scale of 1:50,000 has already been completed and new maps are being produced. The work is supervised by the Institute of Soil Science and Cultivation of Plants and is being performed by the regional offices of Geodesy and Agricultural Instalations.

In recent years, interest has grown rapidly in the problems of the devastation and protection of the natural environment. A map of soil degradation at a scale of 1:500,000 was published by the Institute of Environment Protection. The Institute also produced maps of ecological structure and land use at a scale of 1:10,000 for selected industrial areas in Poland.

Many regions of Poland (mainly voivodeships) have regional atlases produced in cooperation with university centres. They present the natural environment of the region as well as the social and economic phenomena occurring there. The most interesting atlas of this kind produced recently is the *Atlas of the Tatra Mountains National Park*. It contains 63 maps (including 26 maps at the main scale of 1:50,000) on different subjects. Besides thematic maps, the atlas has 14 sheets of topographic maps of the Polish part of the Tatra Mountains.

The most specialised cartographic production organisation is the

Eugeniusz Romer State Cartographical Publishing House. It is responsible for providing the market with maps and atlases for general use and for educational needs. Of particular interest is their new generation of maps of towns with detailed topography and enriched tourist information. Another important activity of the Publishing House is the recent production of tactual maps. The first two maps of Poland at a scale of 1:1,500,000 have been published and are the first of a series. Another recent production is a new atlas of the world for general use. It consists of over 400 maps.

CARTOGRAPHIC EDUCATION

Training of cartographers is carried out mainly at the university level. Four universities are engaged in cartographic education:

—Warsaw Technical University – Cartographic Department in the Faculty of Geodesy and Cartography,
—Warsaw University – Chair of Cartography in the Faculty of Geography and Regional Studies,
—Wroclaw University – Cartographic Department in the Faculty of Natural Sciences,
—M. Curie-Sklodowska University in Lublin – Cartographic Department at the Institute of Earth Sciences.

The first of these educational centres trains engineers of geodetic cartography whereas, in the last three, cartography is a specialisation within geographic studies. Each of the centres produces 7 to 10 Masters of Science graduates per year, except at M. Curie-Sklodowska University where the number of Masters is smaller (about 4 to 5 persons). In total, therefore, about 30 graduates in cartography leave universities every year. Besides the 'regular' cartographic studies, there are three more geographic faculties in Cracow, Gdansk and Poznan where teams of cartographers give lectures to geography and geology students.

At the secondary school level, technicians in topography and cartographic reproduction are trained at geodetic technical schools. It is also possible for alumnae of general secondary schools to be trained in the same subjects in a period of two years.

INTERNATIONAL CO-OPERATION

Participation of Polish cartographers in international cooperation can be seen in the International Cartographic Association, the International

Geographical Union (IGU) and many United Nations agencies. From the very beginning of ICA activity, cartographers from Poland took part in the Association's work. Two Poles have filled the role of Vice-President at ICA. Professor Lech Ratajski, in the period of his term (1972–1977) as Vice-President was also chairman of the Commission on Communication in Cartography. Professor Andrzej Ciolkosz (Vice-President 1977–1984) chaired the Commission on Thematic Mapping with the Aid of Satellite Imagery. In 1982, Poland organised the 11th International Cartographic Conference in Warsaw. Many Polish scientists and practitioners actively took part in the work of several commissions and working groups of the ICA.

Some Polish cartographers are active in the field of co-operation with IGU, especially in subjects connected to cartography and geography. Several commissions of IGU contain members from Poland. Polish experts are also active members of the United Nations Group of Experts for the Standardisation of Geographic Names.

The Head Office of Geodesy and Cartography and some of its subordinate institutions (mainly the Institute of Geodesy and Cartography and the Geodetic and Cartographic Data Processing Centre) take part in international co-operation between socialist countries in the field of cartographic research. This covers, among other topics, the revised edition of the World Map at the scale of 1:2,500,000, publishing the *Atlas of the Council of Mutual Economic Aid*, automation in mapping and comprehensive use of satellite imagery in cartography.

PERSPECTIVES ON NEW DEVELOPMENTS

The Head Office of Geodesy and Cartography plans to produce a new edition of the topographic map at a scale of 1:25,000. It is assumed that computer technology will be used for preparation of the edition. The Geodetic and Cartographic Data Processing Centre (GCDPC) has proposed the concept of digital technology and, in the next few years, work will be focused on installing hardware and development of the software needed. At GCDPC, work on the digital base map is also being carried on. The system will be used by regional geodetic and cartographic enterprises for the purposes of maintaining and up-dating the base map.

Both of the above activities are regarded as important parts of the national land information system (LIS), of which parts are being designed at the GCDPC. Thus the structure of data bases of the topographic map and of the base map are oriented towards incorporation within an information system, not solely to mapping processes alone. LIS in Poland will function at two levels, local and regional. Mutual exchange of information between both levels will be ensured.

Recent research work on the development of a Geographical Information System (GIS) to meet the needs of environmental protection has been carried out at the Geodetic and Cartographic Data Processing Centre, in co-operation with the Institute of Geodesy and Cartography. The system will be based on micro-computer equipment and will collect data from existing maps, remote sensing and other sources. The users of the system will be able to use it as a tool in the processes of regional planning and management of the environment. At the Remote Sensing Centre of the Institute of Geodesy and Cartography, work on a digital image processing system and leading to digital mapping is being developed further. The GCDPC is also involved in the project.

University centres will soon join the current application of computer technology to cartography. Micro-computer equipment is widely available in Poland and, in our opinion, is the best way of proliferating new technological and methodological approaches to cartography.

The application of computer technology is being considered in the production of a new edition of the national atlas called *The Atlas of the Polish People's Republic*. The Atlas will be published by three institutions: the Head Office of Geodesy and Cartography, the Institute of Geography and Spatial Organisation of the Polish Academy of Sciences (IGSO) and the Eugeniusz Romer State Cartographical Publishing House. It is anticipated that publication of the atlas will take place successively in nine thematic parts during the period 1989–1995. The work is presently being carried out mainly by the Cartographic Laboratory of the IGSO.

The propagation of computer techniques in cartography and geography will develop theoretical and methodological research work at universities and research institutes. In the period between 1986–1990, many governmental research programmes are being devoted to geographic information and systems for manipulating spatially-oriented data for scientific analyses, planning, management and environmental protection. The state of cartography in the next decade will depend to a great degree on the results of these programmes.

REFERENCES

Anon. (1987), *Cartographic Activities in Poland 1984–1986. National Report.* Polish National Committee for ICA.

Baranowski, M. and W. Grygorenko (1974), Proba obiektywnego doboru osiedli na mapie z zastosowaniem maszyny cyfrowej. *Polish Cartographic Review*, 6, 4, 149–55.

Baranowski, M. (1980), KARTEM – informatyczny system kartograficznej prezentacji numerycznej danych statystycznych. *Proceedings of Automatyzacja procesow kartograficznych*, Lublin, Poland, 87–92.

Baranowski, M. (1984), Prace badawcze w zakresie utworzenia numerycznego modelu rzezby terenu kraju. *Geodetic Review* 56, 2–4.

Golaski, J. (1982), Interdisciplinary cooperation problems in the standardisation of toponyms on topographic maps. *Geodesy and Cartography*, 31, 1, 85–92.

Golaski, J. (1984), Wokol definicji przekazu kartograficznego. *Polish Cartographic Review*, 16, 4, 172–76.

Grygorenko, W. (1973), Liczbowe kryteria oceny wartosci obrazu kartograficznego. *Polish Cartographic Review*, 5, 3, 117–26.

Grygorenko, W. (1974), Struktualna interpretacja tresci mapy i automatowa grafika maszynowa. *Polish Cartographic Review*, 6, 2, 66–76.

Grygorenko, W. and H. Libura (1977), Przyklad automatowego opracowania linii brzegowej. *Polish Cartographic Review*, 9, 4, 166–75.

Grygorenko, W. (1982), Cybernetyczny model przekazu kartograficznego. *Polish Cartographic Review*, 14, 2, 67–71.

Grygorenko, W. (1984), Ewolucja pogladow na temat roli i funkcji kartografii. *Polish Cartographic Review*, 16, 2, 53–61.

HOGC (1982), *The Polish Cartography*. Head Office of Geodesy and Cartography.

HOGC (1987), *The Polish Cartography*. Head Office of Geodesy and Cartography.

Moscibroda, J. (1975), Roswoj pogladow na metode izarytmiczna oraz jej zastosowan w kartografii ludnosciowej i gospodarczej. *Polish Cartographic Review*, 7, 2, 55–65.

Moscibroda, J. (1978), Podstawowe problemy opracowania map izarytmicznych. *Biuletyn Lubelskiego Towarzystwa Naukowego – Geografia*, 20, 1, 31–9.

Ostrowski, Janusz and K. Podlacha (1980), Automatyzacja procesu sporzadzania map tematycznych na przykladzie systemu PROMEL. *Proceedings of Automatyzacja procesow kartograficznych*, Lublin, Poland, 32–40.

Ostrowski, Jerzy and J. Paslawski (1982), Ewolucja pogladow prof. Lecha Ratajskiego na istote kartografii w opinii prof. K.A. Saliszczewa. *Polish Cartographic Review*, 14, 4, 175–77.

Paslawski, J. (1980), Graficzno-statystyczne sposoby wyznaczania przedzialow klasowych kartogramu. *Polish Cartographic Review*, 12,4, 149–59.

Paslawski, J. (1982), O konstrukcji objasnien kartogramow i map izoliniowych. *Polish Cartographic Review*, 14,3, 114–22.

Podlacha, K. (1983), Jednolita siec pol podstawowych jako uklad odniesien przestrzennych do kodowania informacji w systemie PROMEL. *Prace IGik*, 30, 1, 61–78.

Podlacha, K. (1986), Technologiczne aspekty opracowania map fotograficznych na podstawie obrazow satelitarnych. *Polish Cartographic Review*, 18, 1, 1–10.

Ratajski, L. (1967), Phenomenes des points de generalisation. *International Jahrbuch fur Kartografie*. 7, 143–52.

Ratajski, L. (1970), Kartologia. *Polish Cartographic Review*, 2, 3, 97–110.
Ratajski, L. (1971), Zasady logiczno-semiotyczne uporzadkowania i standaryzcji znakow kartograficznuch. *Polish Cartographic Review*, 3, 3, 106–16, 156–66.
Ratajski, L. (1973), Rozwazania o generalisacji kartograficznej. *Polish Cartographic Review*, 5, 2, 49–55 and 5, 3, 103–10.
Ratajski, L. (1975), Przypadki skrajne w metodzie izarytm teoretycznych. *Polish Cartographic Review*, 7, 3, 100–7.
Ratajski, L. (1976), Pewne aspekty gramatyki jezyka mapy. *Polish Cartographic Review*, 8, 2, 49–61.
Szewczuk, J. (1978), Pole odniesienia. *Proceedings of Materialy Ogolnopolskiej Konferencji Kartograficznej, vol. 6 Problemy map spoleczno-gospodarzych, 37–43.*

THE PRESENT AND FUTURE CARTOGRAPHY IN FRANCE

Jacques Bertin

Cartography has made a great deal of progress in the past few years. It is constantly expanding. It now involves better skills, employs more qualified people and is better taught. Nevertheless, the situation is far from being satisfactory. One has only to read erudite books, skim through an atlas, peruse a scientific magazine, work with researchers or merely look at a map of a city in which one has just arrived in order to be struck by the widespread ignorance of graphic and cartographic language and by the presence of the most rudimentary errors.

CARTOGRAPHIC AND GRAPHIC ERRORS

Errors in graphic language have been perpetrated in the map dubbed the 'synthesis map', which is in reality nothing but a superimposition or combination of data. But a combination of what? It has no legend! This example was taken from an official, national report. Other errors in graphic language are commonplace, such as in the set of four maps transcribing quantitative series with the use of completely disorganised frameworks. Yet this example was taken from a prestigious scientific magazine. The same magazine contains an example of a careless mistake: the graticule is shown reversed and hence the map apparently demonstrates the existence of off-shore boreholes in the Sahara and Tibet. There have also been errors in comprehension, such as when UNESCO had to order the re-drafting of a disastrous ordination analysis, produced after much work, when 300,000 copies had already been circulated.

Geographical nonsense can be seen in a beautifully coloured and extremely detailed map that shows 'the Long March' of Mao Tse Tung, describing his famous trek through the 'great marshy steppe', passing along the ridges of the mountains. Does the draftsman only know how to read the romanticised commentary? It seems as if the cartographer is completely oblivious to what constitutes military expertise and the military value of different types of terrain.

There have also been examples of errors in method when the researcher ignores, or wishes to ignore, the reason for mapping the geographic distribution, such as the distribution of *all* the sites known of a particular civilisation. This information is essential in order to make continuous geographical and historical comparisons.

Finally, there are many other errors in cartographic method which have the most serious consequences. These occur when it is made obvious through the researcher's presentations and publications that he has ignored the various mathematical, graphical and cartographical methods available today. With these, one can compare, classify and group several maps or graphics in order to draw conclusions – rather than just combining them.

It is obvious from all this that the problem is located not only at the draftsman's level. With each additional person – compiler, editor and map author – there is an increase in the degree of incompetence. This situation is evolving. But those whose task it has been to study and improve maps and cartography cannot help but see these events as failures and are therefore searching for the causes of these errors and for those responsible for them.

So far as education is concerned, an International Symposium, 'Education et Cartographie', held in Paris in 1985, illicited some useful information (EetC 1985). It made inquiries into the school and university environments. The results were very instructive. It turns out that:

—the usefulness of cartography is not unanimously accepted by the teachers.
—no teaching on the use of maps or with the use of maps as aids is officially sanctioned – at least not in France and unless it is used (as it very rarely is) in teaching geography.
—the teachers themselves have not received appropriate graphical education and they virtually ignore publications relating to the use of graphics.
—maps are generally considered visual aids and are rarely used as an element of a thought-provoking and decision-making process; even when they are used in this way, it is usually only to pinpoint topographical locations.

THE PEOPLE IN CARTOGRAPHY

The advent of the computer seems to have altered the use of maps. Thanks to the automation of cartography, the use of maps is perceived to have become a helpful exercise. Unfortunately it still attracts far more students than it does teachers. Results such as those quoted seem to indicate what needs to be done. They also show that it is quite presumptuous to ask those who use the maps to express their needs and give their opinions about the drafting of maps, when they have had no formal training on how the judicious use of maps and graphics could aid them.

Despite the change of attitudes brought about by use of the computer, cartography has had a hard time proving itself as a valuable instrument to be used for decisions as well as for communication. There are, of course, some experts who are available to give advice. But who are they? The answer is most easily found in the 1987 report by Le Conseil National de l'Information Géographique (LCNIG 1987). This report demonstrates that these map experts are very few in number and can only be found in a very small number of highly specialised fields. The report goes on to compile a list of the large French mapping operations, including both those that are presently in existence and those predicted for the future. As well as summarising the machine-readable lists of information on geodesy and surveying (IGN), geology (BRGM), cadastre and SPOT images, the report summarises the available area measurements, orography, road patterns, basic scale maps of 1/100,000 and interpretations of SPOT imagery. It also cites information held on forests, land use, pedology, geology and oceanography. Apart from these examples, the official geographic information in France provides us only with topographic information.

The majority of the members of cartographic committees are drawn from the fields listed above and this is where the financial and technical resources are concentrated. The latter are always in short supply yet are huge compared to the resources allocated to general thematic cartography – which is infinitely dispersed amongst different organisations due to its possession of a universal language. Without the fundamental knowledge of topography, there would certainly be no such thing as cartography. But the feeling in France is that, with only a few exceptions, there is no cartography apart from the study of topography. Or, in the words of a well known geographer: can the statistical map really be considered a map?

Is the micro-computer likely to change things? Is it likely to make the use of graphs and thematic maps applicable to all fields and enable decision-makers to benefit from the indisputable values of graphic representation? This question was the theme of a recent symposium dealing with micro-computer programs for cartography (AAMI 1987). There is an abundance of programs, but what are they used for? The response is almost always the same: to make a map! Everything above suggests that the map is indeed seen as an end in itself. In the thoughts and notes of the computer scientists, the map is only rarely used as a part of a decision-making process. The comparison of several maps on the same screen, the interrelation of data and synthesis maps, as well as the division of the results into classes, are rarely found. Programs for producing combinations of Tables – Graphics – Maps are almost non-existent, and trichromatic synthesis is completely absent.

There is a difficult task at hand. At the very least it is necessary to:

—go beyond the topographical 'tunnel vision' that constitutes the still almost exclusive conventions of official French cartography.

—give teachers the information and the necessary means for them to be able to utilise the power of graphical and cartographical methods in the development of thought and in decision-making processes.

—give system builders and programmers of computers an objective other than that of producing 3-D perspectives (a technique already some 400 years old), and prove that the real problem of tomorrow will be that of the 'multiple-dimension' decision. This may well be resolved only by the combined use of data analysis and modern graphics.

REFERENCES

EetC (1985), Explanation given at the international colloquium on 'Education et Cartographie', held in Paris in September 1985 and published in *Le Bulletin du Comité Français de Cartographie*, no. 105, 106, 107.

LCNIG (1987), Report by Le Conseil National de l'Information Géographique, Paris 1987.

AAMI (1987), Réunion de L'Association Aménagement et Micro-Informatique, dealing with Computerised Thematic Cartography. Paris, April 1987.

THE NATIONAL CARTOGRAPHIC PERSPECTIVES IN MEXICO

Nestor Duch-Gary

INTRODUCTION

The Mexican national cartographic series at scales of 1:50,000 and less are produced by the National Cartographic Institute of the General Directorate for Geography of the National Institute of Statistics, Geography and Informatics. The objective is that these maps should cover the entire national territory and be accessible to everyone who wishes to acquire them. But there are many other aspects of cartographic production which must be described to give an adequate, if brief, description of the situation in cartography in Mexico.

The perspectives of national cartographic development are determined, in my opinion, by the interaction of a number of different elements. The most significant elements are the genesis and the current status of our national cartography, the economic situation of our country and the technological development of and need for cartographic information.

I have tried therefore to organise this chapter to take account of the interactions of these elements. Thus, in the first part of the chapter, a brief description is given of the historical evolution of cartography in Mexico. This is followed by a description of the present situation and, finally, given the economic situation and the priority requirements for cartographic information in my country, I comment on the Mexican perspectives resulting from the interaction of the elements.

THE HISTORICAL EVOLUTION OF CARTOGRAPHY IN MEXICO

Don Martin Cortes was the son of the Spanish conqueror Hernan Cortes and his native woman Dona Marina. He was, therefore, one of the first descendants of the Spanish and aboriginal blend that later would be the origin of Mexican nationality. He wrote a study on maritime navigation and he is reputed to have proposed a cartographic projection that could represent, with straight lines, the ship's course. I begin this outline of cartographic evolution in Mexico by pointing out the intuition of a 'Mexican' about the importance of a cartographic projection which was later developed by Mercator and for which he achieved lasting renown.

During the colonial period, there was intense cartographic activity which was basically oriented to navigation. Of course, during this epoch the Europeans were engaged in the making of charts. These Europeans were mainly Spaniards, who surveyed the coasts of New Spain. The cartographic works of Francisco de Gary, made in 1521 along the coasts of the Gulf of Mexico, are worthy of special mention. In 1527 and 1529, charts of this region were produced by Diego de Riviera and published by order from Emperor Charles V. In 1540, the Pacific coast was explored. One expedition, commanded by Hernando de Alarcon, reached the Gulf of California. Domingo del Castillo, first mate of this expedition, produced a chart of that journey considered to be the most ancient cartographic product relating to the occidental coast of New Spain.

Mexican cartography properly considered began with Don Carlos de Siguenza y Gongora, author of a *General Map of New Spain*. This map was considered the best of the colony up until the end of the 18th century. Siguenza's map was replaced by the *New Geographic Map of Northern America*, belonging to the Vice-royalty of Mexico, produced by Alzate y Ramirez and printed in Paris in 1775. During the same period were made the map of Joaquin Valadez, a Mexican scientist (made to describe the position of the most noteworthy mines of the country), and the *Manuscript Map of the Whole Kingdom of New Spain*, the work of Antonio Foncada y Plaza. Both works are mentioned by Humboldt in his *Political Essay on the Kingdom of New Spain*, who described them as 'cartographic documents of positive value'.

In 1856, the *Geographic Statistical and Historical Atlas* by Antonio Garcia Cubas was published. Two general charts of the Republic at a scale of 1:2,000,000 formed part of the atlas. To produce these two charts,

Garcia Cubas extracted material from those published by Humboldt in 1811 in Paris who, in turn, had extracted material from Alzate's chart. In 1877, the Geographic Exploration Commission was established, the first institution in charge of the national cartographic endeavour. During its 31 years of life, it observed 800 astronomic positions and established more than 200,000 kilometres of traverse. It published 197 maps at a scale of 1:100,000 and data corresponding to 300 maps and several charts of different states of the country.

The Commission, however, suffered much criticism for its procedures, mainly for not giving enough attention to the geodetic net and for using only astronomical methods. To remedy these defects, the Mexican Geodetic Commission was created in 1899. It had also, among its first tasks, collaboration with the United States of America to measure a meridian arc. The initial point was located on the coast of Oaxaca and the measurements were made along a 45 degree arc of the 98 degree West-of-Greenwich meridian through Mexico, the United States and Canada – no less than one eighth of the world's circumference. This Commission was almost exclusively engaged in geodesy rather than in cartography. The Commission was followed by the Directorship of Geographic and Climatologic Studies, created by President Carranza in 1915. This Directorship was responsible for the synthesis of geodetic and cartographic interests and initiated the era of modern national cartography in Mexico.

Around 1916, the cartographic coverage of the country was represented by a map of the Republic at a scale of 1:2,000,000; by a chart of the Federal District, including the City of Mexico; and by sheets and charts produced by the Geographical Exploration Commission of individual states. In 1925, however, the Directive Council of Topographic Surveys in the Republic was created for the purpose of establishing homogeneous standards and procedures in order to unify the specifications used up to that time. Two years later and through the initiative of this council, a 'Project to form a Pan-American Geographic Institute' was proposed for the purpose of extending the standardisation and homogenisation criteria in map-making to the whole continent. In the following year, during a conference held in Havana, Cuba, all the countries of the continent adopted – with few modifications – this Mexican proposal. Thus was born the Pan-American Institute of Geography and History (PAIGH) which has largely fulfilled its original functions up to the present day.

Space allows only a brief comment on the work of the General

Directorship of Geography and Meteorology which led to a significant advance in national cartographic production. In 1938, the Military Cartographic Commission began its activities with the aim of producing the country's military charts. This Commission, today called the General Directorship of Military Cartography and which comes under the National Defense Secretariat, continues to improve its cartographic products. In parallel, the General Directorship of Oceanography and Maritime Signaling has the responsibility for producing nautical charts and has also achieved substantial production in that area.

To conclude, we should note the names of the following individuals who have all made notable contributions to Mexican cartography in the last 150 years: Colonel of Engineers Agustin Diaz, Engineer Angel Anguiano, Engineer Pastor Rovaix, Engineer Pedro C. Sanchez, Doctor Sotero Prieto, Engineer Carlos Bazan, Engineer Isidro Diaz, Engineer Manuel Madina, Professor Carlos Martinez Becerril, Engineer Horacio Vazquez Glumer and General Orozco Camacho. The current generation of Mexican cartographers owes them much.

THE PRESENT SITUATION

Although the Directorship of Geography and Meteorology continued to operate, its cartographic activity declined substantially. At the same time, other governmental and private organisations also produced charts basically oriented to solving specific and local problems. This situation prevailed until 1968. In that year, promoted with great tenacity by Engineer Juan B. Puig de la Parra, the 'Commission to Study the National Territory' was created. This revived, with a new impetus and a very modern vision for the time, the old purposes of unity, normalisation and homogeneity of national cartography. The Commission initiated the cartographic/topographic surveys of the country at a scale of 1:50,000. Thematic cartography was incorporated in its work and it initiated the production of geological charts, together with maps of soil use, climate, pedology and of agricultural potential. In a short period, the Commission acquired very substantial vertical integration. It established its own fleet of aircraft to obtain aerial photographs and created its own print shop. It developed geodesy and took advantage of all the data previously established to complete a geodetic net to control cartography at a scale of 1:50,000.

After 19 years of uninterrupted work, the General Directorship of

Geography (now named the Commission to Study the National Territory) merged with the National Institute of Statistics, Geography and Informatics and the scope of the country's cartographic position can now be characterised as follows. There is a horizontal net of land control formed by triangulation and trilateration arcs, first order geodetic traverses and 40-pass Doppler stations. The densifying nets are formed by circuits of second order geodetic traverses and 12-pass Doppler stations and other points of this net are monumented and total about 7,800 vertices.

The vertical net is formed by circuits of first and second order levelling, related to mean sea level and adjusted to sea gauges placed in the principal ports of the country. There are about 15,475 bench marks in this net. The net is densified by levelling of lower orders established at right angles to the lines of photographic flight. There are 12,550 bench marks on those lines. The tide gauge stations, which are controlled by the Geophysical Institute of the National Autonomous University of Mexico, amount to 13 stations in the Pacific Ocean and eight on the Gulf of Mexico.

So far as basic topographic cartography is concerned, the country is now covered completely by maps at scales of 1:1,000,000, 1:250,000 and 1:50,000. Completion of the last of these involved the production of 2,320 map sheets over a period of 15 years. The surveys for thematic cartography started with the scale of 1:50,000 and by 1979 were continued at the scale of 1:250,000. Considering both scales together, the country is also totally covered by geological charts; the remaining thematic topics are programmed for completion by the end of 1988.

The General Geographic Directorate has photographic coverage of the entire continental territory at a scale of 1:75,000 and also holds coverage of the more rapidly changing and economically important regions at a scale of 1:37,500. In 1981, the Directorate designed and implemented the NSAP (National System of Aerial Photography) which permits a technically optimum operation of the flying capacity. NSAP output is also produced at scales of 1:75,000 and 1:37,500.

Other nationally important work is produced by the General Geographic Directorate and includes the chart of the Patrimonial Sea and Economically Exclusive Zones, the Geoidal Chart at a scale of 1:5,000,000 and the Bathymetric Chart of the Country's Coasts at a scale of 1:1,000,000. The same Directorate has published three atlases: *The Physical Ambient Atlas*, at a scale of 1:1,000,000; the *Chart of Mexico*, at a scale of 1:250,000; and an *Atlas of Historical Cartography*.

There are, of course, other governmental, private and academic institutions in the country which produce maps. Among the first are the
Hydrologic Commission of the Valley of Mexico, the Secretariat of
Agrarian Reform, the Secretariat of Communications and Transportation, the Secretariat of Urban Development and Ecology, PEMEX,
the National Register of Electors, the Federal District's Treasury and
the Secretariat of Agriculture and Hydraulic Resources. In most cases,
these institutions produce (sometimes through private organisations)
cartographic documents oriented to very specific problems, with very
local coverage. Generally, these documents are not reproduced and the
general public has no real access to them. However, in the author's
opinion, three agencies deserve special mention. These are the Federal
District Treasury, which has made significant advances in urban cartography and in creating digital cartographic systems aimed at urban
planners, the Secretariat of Urban Development and Ecology which has
developed a very efficient system to produce ortho-photomaps of small
and medium cities at a national level, and the Secretariat of Agrarian
Reform which has introduced a digital system to produce maps related
to the rural cadastre of the country. In the last case, there have been
certain controversies about the precision and cartographic rigour of the
maps but it is certainly an innovation worthy of attention.

The private institutions engaged in cartography mainly dedicate their
efforts to developing topographic cartography at a very large scale
associated with engineering projects or with the construction of various
infrastructure projects. As far as the author is aware, only a few have
analytical equipment for photogrammetric restitution and have obtained
contracts abroad. One such private organisation is a medium-sized
Mexican enterprise which, notwithstanding its relative youth, has started
to develop certain new cartographic products of very good quality and
which are commercially distributed to the public. This institution has
been characterised by continuous innovation and by a wide range of
activities. In addition, two Mexican editorial companies have produced
atlases of a classic type oriented to students on a middle level whilst the
Geographic Institute of the National University of Mexico has been
heavily involved in the conception and design of a new national atlas,
including many new concepts.

MEXICAN PERSPECTIVES

From the point of view of complete national coverage, the country's
mapping at the medium scales of 1:50,000 and 1:250,000 and the desire

for limited large scale production has, for the most part, been met. Although considerable discussion has taken place, we are not yet in a position to make a decision on whether the huge effort to obtain larger scale coverage is justified. One attractive option is to consolidate and up-date the work already accomplished. Priorities in the planning of national cartographic development are the conservation, gradual integration and re-adjustment of the national geodetic net, creating an archive from the national system of aerial photographs, up-dating of the topographic series at scales of 1:1,000,000, 1:250,000 and 1:50,000 and the completion of the thematic series.

We must also take into account the dynamic changes now taking place in cartographic technology. The accelerating advance in the sophistication of equipment and instruments used in different stages of cartographic production, including its growing digital nature and the development of satellite technology, remote sensing and geodetic positioning, need attention by management in the immediate future. We are also very much aware of the growing influence of developments in theoretical cartography in regard to the conception and design of more rigorous maps which should be much easier to read and more clearly intelligible to a larger sector of public users. Finally, the financial context is of significant importance since Mexico is immersed in an economic crisis at present. This economic situation severely limits the availability of substantial investments to acquire equipment and instruments more technologically advanced than those available at present and makes the supply of imported materials difficult and expensive.

From another perspective, the size of the Mexican territory, the dynamics of geographical changes in many of its regions and the complexity of its geography requires – more urgently with each map series – precise cartographic documents. These must be able to provide necessary information and which have been up-dated with the latest statistics; in consequence, they could then be used properly by a larger portion of the population.

All these different elements are inter-related. The author believes that from these inter-relationships arise a series of concurrent trends for the future which will impact upon the development of national cartography.

The maintenance of what has already been done, the improvement of its quality and the completion of cartographic series already initiated are clear tasks for Mexico in the near future. These tasks should be approached using all those new technologies which contribute to improvements in quality, to making the work easier, and to speed production at a lower cost. New technologies should be introduced gradually, however, according to the economic conditions of the country and must be paralleled by the necessary investments in the training and instruction of personnel. The introduction of new techniques should conform to a global plan, not one that is strictly defined and rigorous *a priori*, but one that is flexible, feasible and incremental. One further task in the planning of national cartography is the assembly of integrated geographical information about the country (especially that held in digital form). Mapping produced from these data bases needs to be tailored to suit the needs of planning of the land.

In our country, the profession of cartographer as such does not exist at medium to higher levels in administration or education. Mexican cartographers have been educated abroad only in small numbers. Others have been trained in the General Geographic Directorate itself. In this latter case, their education is mostly related to the practical needs of production. This situation constrains reflection and theoretical discussion and hence discourages change and innovation. In the author's view, it creates a marked tendency to insist on the repetition of procedures and models already known and accepted. Further developments in cartographic education are needed.

The need to strengthen already incipient 'cartographic policies' is becoming ever more evident within the General Geographic Directorate. There is a need, for example, for an agreement about procedures and methods of cartographic surveys which permit compatibility of surveys performed by different organisations, both public and private. In 1984, the minimum standards for geodetic surveys were officially published for endorsement. The minimum standards to produce topographic maps at medium scales will appear before long. The definition of clear standards such as these is a clear and important task for the future development of national cartography.

CONCLUSIONS

If I had to characterise national cartography in Mexico, I would say that the present situation is one of transition. Great efforts have already been made to obtain a basic national coverage but we must now go through a stage where the emphasis should be put on the qualitative aspects of our cartographic coverage. We must consolidate, improve and complement what we have produced to date. This change must be made within the context of the adoption of new techniques in a gradual and very selective way. My final proposition is for the creation of systems of geographic information which are modern, opportune and organised in such a way as to support the challenges facing national development. These challenges require increased knowledge and the better use of our territorial space.

THE STATE OF CARTOGRAPHY IN THE FEDERAL REPUBLIC OF GERMANY

Ulrich Freitag

THE PAST AND ITS INFLUENCE ON THE PRESENT

As a state, the Federal Republic of Germany is young; so are her cartographic institutions and enterprises. She was officially founded in September 1949 as a result of political and economic integration of the American, British and French (western) occupation zones of Germany and of her separation from the Soviet (eastern) occupation zone, which subsequently became the German Democratic Republic in October 1949. The four occupation sectors of Berlin developed in a similar way: the western sectors formed the state of Berlin with special relations to the West German Republic whilst the eastern sector of Berlin became the capital of the East German Republic.

The development of the two German republics has been influenced by two opposing tendencies. The first is their political, military and economic segregation due to their inclusion into different, opposing treaty organisations under the dominance of the world powers; the second tendency is their cultural integration due to their common history, language and culture. The strength of the cultural ties may readily be illustrated. Switzerland became politically independent of the First German Empire as early as 1499 and Austria was excluded from the formation of the Second German Empire in 1870. Yet, in spite of the political differences – manifested even by opposition and hostility between these political units – the common German language facilitated the exchange of concepts and experience, sciences and techniques within the cultural entity. The development and state of cartography in the Federal Republic of Germany must be seen in this context.

For the development of official cartography, one factor became of dominant importance: the political organisation of the republic. In central Europe, the diversity of the geographic environment favoured the formation of regional territories rather than that of centralised empires. This tendency persisted even when, in the last 120 years, two German empires were established and destroyed in two World Wars. As a consequence of all this, the newly formed Federal Republic of Germany tries to preserve a careful balance between the central federal government and the newly organised, strong regional state governments. Official cartography reflects this organisational structure. As a result, there is a diversity of locations of federal and state departments with cartographic offices in the federal capital, in the state capitals and in other cities (Fig. 5). In addition, many cities developed important commune-level survey and mapping branches.

For the development of commercial cartography, two factors were of importance: the destruction of commercial establishments and valuable materials during the last war, and the economic and political division of Germany. This division resulted in the development of new enterprises in new locations and added to the diversity of locations of cartographic production; in some cases, the division led to the duplication of carto-graphic enterprises in the old and in new locations. Moreover, com-mercial cartography was long produced mainly for the German-speaking market, but the competition of a growing number of enterprises in a stable market led to two developments: the concentration into bigger enterprises and the specialisation in specific cartographic products. Only recently have international co-operation and business connections directed the attention of German map producers to the potential interna-tional market.

Official and commercial cartography both have a long and strong tradition in Germany. Maps were produced as a means of providing information and facilitating actions of a geographical or spatial nature but they were also kept and studied as documents of history and cultural heritage. Therefore technical and scientific aspects of cartography were taught in many places – at offices, commercial enterprises, technical schools and universities. The first textbook on scientific cartography was published in German about 60 years ago and the first courses for cartographic engineers were launched in Germany 50 years ago. Based on this tradition and on the rapidly increasing demand for qualified

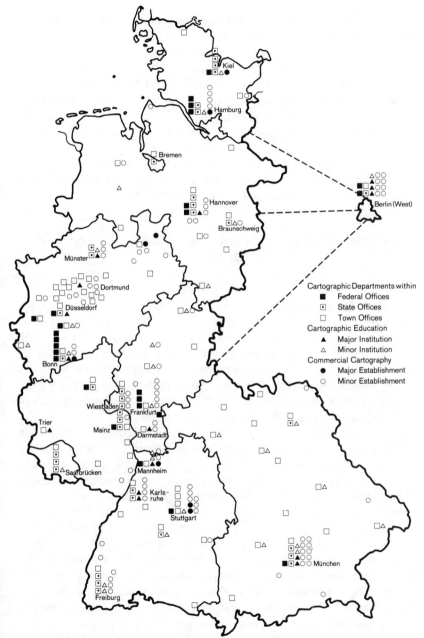

Fig. 5. Federal Republic of Germany: cartographic offices and establishments.

cartographic personnel, technical and scientific training and education in cartography developed in various new locations, influenced by state policies and commercial considerations (see Fig. 5). Also arising from this tradition, a German Society of Cartography was formed 50 years ago and re-established after the Second World War in Bielefeld as a meeting forum and as a professional organisation of cartographers of all levels of education and interest.

Superficially, therefore, the present state of cartography in the Federal Republic of Germany can be seen as a young discipline in many new locations within the framework of the federal organisation of the young republic. But to understand German cartography, the influence of former locations and of older traditions must be considered.

THE PRESENT: CARTOGRAPHIC ORGANISATIONS AND PRODUCTION

Sources of Information

The state of cartography of the Federal Republic of Germany is regularly reported in various publications. The annual production of maps is listed in a variety of map catalogues of the federal, state and city mapping departments. Commercial cartography products are listed in catalogues at the annual book fair in Frankfurt. The output of cartographic papers is listed in the annual volume of the international documentation of cartographic literature *Bibliographia Cartographica*.

At regular intervals, national reports on the cartographic activities in particular periods are submitted by German cartographers to the General Assembly of the International Cartographic Association, as well as to other meetings of international organisations. Two special sources of information are the outstanding books which were published by the German Society of Cartography on the occasions of the 19th German Cartographic Conference in 1979 in Vienna and of the 33rd German Cartographic Conference in 1984 in Fellbach; these conferences were held jointly with the cartographic societies of Austria and Switzerland. The publications were entitled *Kartographie der Gegenwart in der Bundesrepublik Deutschland* (*Today's cartography in the Federal Republic of Germany*) and are each a detailed descriptive survey of cartography with a collection of numerous map sections in colour.

Other sources of information on the state of cartography are the journal of the German Society of Cartography *Kartographische Nach-richten* (Cartographic News) which, in 1987, was in its 37th volume, and other geodetic and geographic journals. Another important source of information is provided by the numerous series of cartographic titles published by some official mapping departments and many institutes of cartographic training and research. All of them are listed in the *Bibliographia Cartographica*.

This great variety of sources illustrates the intensive cartographic activities in all parts of the country. But, as most catalogues, reports, books and journals are published only in German, many of them are not known to the international cartographic community. It might be worthwhile for people with less lengthy experience and traditions in cartography to look at them and draw their own conclusions on the accumulated knowledge of German cartographers.

Official Cartography
Official cartography includes all the activities of federal, state, and local government offices. Its main task is the provision of maps to the government at all levels of administration, public service and provision, public security and defense. While formerly the cadastral and topo-graphic map series of the survey departments were adequate tools for many government offices, the recent differentiation and specialisation of these offices to deal with specific topics and tasks has led to the establishment of various specialised mapping departments for monitoring and planning particular spatial topics.

The most important mapping organisations today are the survey and mapping departments, generally attached to the federal and state Min-istries of the Interior. They produce and up-date the topographic map series. The basic topographic map series are at scales of 1:25,000, 1:50,000 and 1:100,000; they are produced and distributed by the 'Lan-desvermessungsamter' (state survey departments). These departments discuss common problems within formalised working groups ('Arbeit-skreis Kartographie der AdV'). These groups worked out an agreement to assign specific areas to individual survey departments. They also agreed on common standards and specifications for each map series. These are revised at appropriate intervals to adapt them to the changes of map production techniques.

The German topographic map series exist at several different scales. The map series at 1:25,000 (TK 25) consists of 2,093 sheets. These sheets existed in various forms before 1949; they were then re-designed and are regularly revised. The map series 1:50,000 (TK 50) in 558 sheets was newly produced between 1957 and 1963; it is the main military map and the most popular series for official and private use. The map series 1:100,000 (TK 100) in 152 sheets was newly produced between 1960 and 1987. The state survey departments are responsible for two more map series. The first of these is the large-scale German Basic Map at 1:5,000 scale (DGK 5) which, when complete, will consist of 75,311 sheets; it is produced as a photomap or as a line map in various forms: the series has not been completed in some states while in others the sheets are regularly revised. The second series is formed by the 44 sheets of the small-scale map series 1:200,000 (TUK 200) which were newly produced between 1958 and 1973 by the federal survey department ('Institut fur Angewandte Geodäsie' or IfAG), following a special agreement with the state survey departments. In contrast, IfAG is fully responsible for the production of the general map at 1:500,000 scale (UK 500) in four sheets, newly produced between 1974 and 1976, and also for the International World Map at 1:1,000,000 scale in two sheets, produced according to the 1962 specifications in 1977. All map series of the survey departments are published in various versions. These include monochrome versions on paper or film as working sheets for many users and polychrome paper versions mainly for the public. All of the map series (except for some sheets of the DGK 5) are line maps, but their revision is mainly based on air photo interpretation. Recently, more and more sheets are compiled to cover specific urban or other geographic areas with the intention to be useful as traffic, hiking or planning maps. Each state has also compiled at least one topographic atlas of its territory, containing sections of topographic maps with adjoining explanatory text; the last topographic atlas was the *Atlas of Berlin*, published in 1987.

In many of the states, the topographic survey departments include cadastral survey departments or act as their supervisor. New cadastral maps are now produced according to generally accepted specifications but, in many areas of Germany, we still find a great diversity of scales, coverage and standards of cadastral maps. One of the major tasks of mapping in the densely populated and intensively used land of Germany is the establishment of a uniform land register. This task can only be achieved by using standard procedures with digital adjustment and digital mapping. The final product will be the 'Automatische Liegensch- aftskarte' or ALK (Automatic Land Register Map). This map is an essential tool of taxation, land consolidation, pipe and cable monitoring, land use planning and construction. It may also form the geodetic base of a general Land Information System (LIS) that is being intensively promoted by many land surveyors.

All of the map series of the survey departments are being used as precise base maps in the monitoring and planning of particular projects by many other government offices. Among these offices are the federal and state offices of geology and mineral resources, the meteorological offices, the hydrological offices, the offices of environmental protection, forestry offices, the land consolidation offices and the offices of federal, state, regional and commune-level planning. The map content and the map graphics of some of the resulting thematic maps are fixed by specific standards. These may be international agreements on terms and graphics; alternatively, they may be federal or state laws like the federal construction law, in conjunction with the state regional planning laws or the federal nature protection law. A map is of judicial value only if it strictly follows the specification which is set by the appropriate law and the attached ordinances and regulations. The discussion of the validity of the 'Bauleitpläne' (Land Development Plans) plays an important role for society, groups and individuals.

The basis of federal, state and regional planning is provided by planning atlases. The federal planning atlas (*Atlas zur Raumentwicklung*) was produced and published in ten volumes between 1976 and 1987 by the Federal Research Institute of Regional Geography and Planning. The planning atlases of the federal states were compiled between 1950 and 1982 by the Academy of Regional Planning and the respective state offices; they contain between 115 to 499 maps on 73 to 170 pages. It should be also be mentioned that various cities developed and published their own city planning atlases, among them the cities of Bochum, Dortmund and Wuppertal.

Other federal or state departments have to survey and to produce their own specific maps. The German Hydrographic Bureau (DHI), for example, is recognised internationally for its hydrographic and navigational charts. Some other departments use their own maps to construct and control traffic and traffic installations: among them are the German Federal Railway, the Federal Bureau of Air Navigation Safety and various road construction departments.

Commercial Cartography

The German *Encyclopaedia of the History of Cartography*, published in Vienna in 1986, points out that commercial cartography has a strong tradition in all German states. German wall maps, reference atlases and scientific maps have long been internationally used and acknowledged. The devastation caused in World War II diminished the capacity of commercial cartographic enterprises considerably, but the rapid general economic growth after 1949 helped commercial cartography to produce traditional and new maps and atlases of a high standard.

At present, three major fields of cartographic production can be distinguished: the production of general geographic maps and atlases, the production of educational maps and atlases and the production of maps and atlases for travel and tourism. The last field developed rapidly after 1950 and is the most competitive and extensive field of cartographic production.

The production of general geographic maps and atlases continues a long tradition connected with the hand atlases of Stieler, Andree and Debes. Several atlases were newly developed purely as world reference atlases, including the *Bertelsmann-Atlas*, the *IRO-Atlas*, the *Herder-Atlas*, the *Meyer-Atlas* and the *Rand McNally/Westermann-Atlas*. In addition, several German editions of foreign atlases competed with these indigenous atlases on the German language market. This resulted in two developments. One result was the improvement of precision and graphics of relief representation and the search for and adoption of an internationally accepted lettering system for place names. The other result was the publication of more and more versions of the same maps, either as cheap versions of atlases with a reduced number of maps or colours or as expensive representative versions with enlarged maps and more gererous pagination. A new development is the production of family atlases. They combine the general reference map with sections of texts, photographs, graphics and tables on the universe, the world, the states and the environment. The text section is often 'glossy and flashy'. Sometimes it contains maps from scientific or educational atlases.

The production of educational maps and atlases also continues a long tradition connected with the school atlases of Kirchhoff and Sydow-Wagner and the wall maps of Sydow and Haak. Justus Perthes continued the wall map tradition and became one of the leading world producers of topographic and thematic wall maps with lettering in Romanic, Arabic, Hebrew and other scripts. In the field of school atlas production, revolutionary changes occurred in the 'seventies. Four new atlases were conceived and produced in one decade. In 1974, the new *Diercke-Weltatlas* was published with 200 map pages in standard A4 format. (A totally re-designed and extended edition of the *Diercke-Weltatlas* will be published in 1988.) In 1975, the *List Großer Weltatlas* was published with 150 larger map pages and many economic maps with pictorial symbols. One year later, the *Alexander Weltatlas* was published; on its 144 pages the newly developed, consistent geographical base maps for all countries and a new arrangement of thematic maps were outstanding features. In 1984, the *Seydlitz Weltatlas* was published; it was the redesigned and extended version of the *Atlas Unsere Welt* of 1978. Consisting of 184 map pages, its main characteristics are the representation of each region on one physical and one economic map. Each atlas was published later in revised editions, as well as in reduced and regional editions. Better than those in any other cartographic product, the changes in the school atlas illustrate the radically altered concepts of the natural environment and of our attitude to our social environment.

The production of maps and atlases for travel and tourism rapidly extended and diversified with the general economic growth and the spread of cars after 1950. The production of road maps and road atlases was directly stimulated and supported by the automobile and petroleum industries. The earliest (and still outstanding) products of this type were the series entitled *General Map of Germany* at 1:200,000 scale in 26 sheets (1957) and the *Great Shell-Atlas*, the first comprehensive road atlas of Germany and Europe with many city maps and tourist information (1960). They served as successful models for many cartographic enterprises. Today, road map series of uniform scale, layout and folding, content and design are the trade marks of several publishing houses. The map scales used are 1:200,000 to 1:400,000 for regional map series and 1:800,000 to 1:1,000,000 for national map series of European and other tourist countries. In addition, these maps form the base for comprehensive road atlases with different map parts, including city maps and tourist information. The volume of the road atlases increased considerably as many more tourist regions became popular and includes the cities of Europe and the Mediterranean. It became cumbersome to carry and use them and difficult to revise them. As a reaction, a new type of road atlas appeared in 1985: a slim volume of two types of road maps of Germany and Europe only, produced in standard A4 format, inexpensive, easy-to-use and with a date on the title page to indicate that it was recently revised. Again, this successful atlas served as a model for several similar, slim road atlases.

Another field of activity is the production of city maps and plans. Although the survey offices of many cities produce official city maps, the city plans of commercial enterprises are often preferred by the public. Commercial city plans are easier to obtain, to handle, to read, and contain more up-to-date traffic information than do most official plans. Moreover, the content and design of commercial city plans does not change from city to city; on the contrary, the standard folding and layout, standard content and graphics were developed as trademarks of publishing houses, best illustrated by the Falk plans.

In recent years, a new type of city plan has become popular with the tourists who used them as attractive souvenirs rather than as a means of orientation: the perspective city plan. The idea was converted into reality for many small towns and city centres (including Amsterdam and New York) by Bollmann Verlag over the period from 1948. Now several architects and artists have followed this example for their own home towns.

Another field of activity is the production of hiking maps. Large scale topographic maps of specific tourist areas with additional information for the tourist were the main products for many small cartographic enterprises. This production is more and more limited by the provision of commercial editions of the official topographic maps and the increase of tourist information on large scale road maps. Therefore some enterprises have tried to produce very specialised maps for particular users – the cyclist, the climber, the photographer, the sporting tourist – or to produce maps of special areas with the support of the regional tourist office. As in the case of urban areas, a new type of cartographic representation became very popular for hiking and tourist districts: the perspective map or panoramic map. Very often these are not only attractive souvenirs but are used as a means of orientation.

Finally, it should be mentioned that a few cartographic enterprises obtained the support of travel agencies to develop special map series of tourist areas, not only of Europe but also of many parts of the world. Small and handy, easy to read and with all the essential data for a region or a country, these provide comprehensive information for the group tourist.

Cartographic Education and Research

For centuries, the professional education and training of cartographers took place within official cartography departments and commercial enterprises. The mobility of manpower and the increasingly higher costs of training have reduced the number of trainees of this type considerably over the last few years. One more reason for this decline is the fundamental changes in map production techniques; they either require more specific technological knowledge and experience in conventional typographic map production or knowledge in the new field of electronic map production. To adapt the existing curriculum to the changing technological environment, a pilot training project for a combined training in both established and new techniques is being carried out at present.

A higher level of cartographic education can be obtained at the polytechnics ('Technische Fachochschule'). After training and four years of study, the students may obtain the diploma of cartographic engineer (Dipl.-Ing., FH). Polytechnics exist in Berlin, Munchen and Karlsruhe. Cartography courses were offered in Berlin as early as 50 years ago, while Karlsruhe started only in 1978. All polytechnics produce numerous cartographic engineers for official and commercial cartography; some variations exist in the training in that certain students may be more interested and trained in cartographic techniques or, alternatively, in cartographic compilation and design. Practical work and research projects of different kinds contribute to the training and specialist knowledge.

Another level of cartographic education can be obtained at the universities. Cartography here is offered as a minor subject, either within geodesy and photogrammetry at the universities of Berlin, Bonn, Hannover, München and Stuttgart, or within geography at the universities of Berlin, Bochum and Trier. The education is less formalised than at the polytechnics. After four to six years of study, the student may obtain the university diploma of survey engineer (Dipl.-Ing.) or geographer (Dipl.- Geogr.) based on a research paper as thesis. He or she may then continue these studies and research and work for a doctorate thesis and degree. All theses are published to document the direction and state of research of the respective cartography department of the university; they appear in the *Bibliographica Cartographica*.

Cartographic research at different levels with different orientation – often with different levels of commitment and results – is carried out at all institutions of higher learning. Most of them concentrate on the development and application of digital mapping, each with a distinct specialisation. Cartographic research is also carried out at several departments of official cartography. An outstanding example is the Federal Survey Institute, with research activities in geodesy, photogrammetry, remote sensing and cartography – a unique combination in Germany. Cartographic research is also supported and carried out by some commercial enterprises in their search for more functional products and means of production. Finally, it is also carried out by departments and enterprises which develop and promote the new techniques of computer-assisted graphic design, although they concentrate on the development of competitive hardware and general software, rather than on the solution of cartographic problems.

The great diversity of cartographic production, education and research requires a forum for the exchange of ideas and experience. Specialised scientific or technical fora are organised by universities, polytechnics, cartographic research departments or vendors of equipment. The annual meeting of the open Working Group on Automated Cartography should, however, be mentioned. General technical and scientific fora are organised by the German Society of Cartography for interested members as working seminars with special topics ('Arbeitskurse Niederdollendorf'). The biggest convention of German cartographers and the most open forum is the German Cartographic Conference; it is held every year in a different city. At this, all working groups of the German Society of Cartography meet to report on their activities, discuss them and develop plans for future work. They concentrate on problems of cartography in Germany, but they also consider the international cartographic problems. The most outstanding speaker to address the German Cartographic Conference several times on international cartographic problems and developments was the long-time Secretary-Treasurer and President of the International Cartographic Association, Professor Dr F.J. Ormeling. His informative, humorous and very personal addresses were all received with great pleasure by the German cartographers.

The generally accepted body of cartographic knowledge is sometimes presented in the form of textbooks or encyclopedias. Both forms of publication provide objective information, but their arrangement, choice of topics and style of presentation are clearly influenced by the individual author. The scientific development of cartography in Germany was strongly influenced by two major types of publications. One type comprises the three outstanding textbooks on thematic cartography, that of the Austrian, E. Arnberger, published in 1966 (554 pages), the text of the German, W. Witt, published in 1967 (383 pages, with a second edition in 1970 of 576 pages), and the textbook of the Swiss, E. Imhof, published in 1972 (360 pages). The second type of publication comprises the various volumes of the *Encyclopedia of Cartography and Related Fields*, initiated by E. Arnberger and I. Kretschmer. Volume I on cartography and topographic maps was published in 1975, Volume II on mountain cartography in 1983 and Volume III on urban cartography in 1987. A *Dictionary of Cartography* was published in 1979 as Volume B and a *Dictionary on the History of Cartography* was published in 1986 as Volume C.

THE FUTURE: TENDENCIES AND TASKS

To outline the future of a discipline can only be speculative. If the speculation is to achieve any degree of accuracy, it should extend current trends and tasks of cartography and take account of the influence of technical innovations and social and economic development in limited fields and special branches of cartography; this is done in the following chapters. The outline of our future has, however, to consider and must be based on a detailed survey of the present state of cartography; this chapter has attempted to describe briefly the essential factors of cartography in the Federal Republic of Germany and to indicate some recent changes which may be interpreted as trends influencing the nature of our future cartography.

CHINA'S CARTOGRAPHY: THE PRESENT SITUATION AND FUTURE PERSPECTIVES

Hu Yuju and Fei Lifan

As is known to all, the basis for surveying and mapping remaining from the Old China was very weak and very disorganised: it was inadequate to satisfy the development needs of New China. Since the founding of the People's Republic, the Chinese government has devoted much attention to the improvement of surveying and mapping in the entire country, and has adopted and implemented the strategic decision to reconstruct completely the surveying and mapping system in China.

Through the joint efforts of various national economic and military departments over the past 30 years, China's astro-geodetic network and elevation system has been established. Furthermore, in the mid-eighties, the total adjustment of the nationwide astro-geodetic network was satisfactorily accomplished : hence the scientific base for China's topographic map series has been assured. Depending on the local situation in different regions, the entire territory has been covered with topographic maps at scales of 1:25,000, 1:50,000 or 1:100,000. Some regions have had their maps revised through a second or even third generation map series. Based on all this, the compilation and publication of nation-wide topographic map series at scales of 1:100,000, 1:200,000, 1:500,000 and 1:1 million have been completed. These map series have become the main foundation for the compilation of China's various maps at smaller scales. Currently, the compilation of the 1:250,000 scale topographic map series, which will be the substitute for those at a scale of 1:200,000, is being undertaken. In the comparatively developed economic and agricultural areas, many topographic maps at a scale of 1:10,000 have been, and continue to be, produced. The task of revising various topographic maps has now been put on the agenda of the surveying and mapping organisations. It was inconceivable in the Old China to carry out basic topographic mapping on such a large scale and in such a systematic way. These recent achievements have met the needs of China's economic development and national defense quite well. They are the material foundation for the developing level of China's contemporary cartography.

The National Bureau of Surveying and Mapping (NBSM) was established in 1956 and is the organisation designated to lead and organise the development of the national survey and mapping undertaking, including production, education and scientific research. Thus, China's surveying and mapping has become a systematic science of great vitality; developments in it occur ever more rapidly and the gap between the advanced international level and China's level has been narrowed. We recognise, however, that still more efforts have to be exerted on a series of problems before we can reach as high a level as the developed world but believe that we have a bright future in this respect. To demonstrate recent progress and future prospects for cartography in our country, we begin with a general description of map-making in the New China.

THE PRESENT SITUATION

Map Production

With the development of more economic production methods, general geographic maps at various small scales and atlases have been compiled and published, derived from a series of topographic maps of the entire nation. Beginning in the nineteen sixties, the publication of a national atlas, a large format physical geographical atlas and large wall maps of China and the world were symbols of the rapid rise in China's cartographic abilities. At present, a new edition of the large national atlas in five volumes is being compiled. It is expected that the first two volumes, the general geographical atlas and the agricultural atlas, will come out in 1990. The other volumes, which consists of physical, economic and historical atlases, are to be compiled and published later.

Mapping is not confined to the national territory. In 1986, the large format (35 cm × 38 cm, 190 pages of maps and text) *Atlas of Africa* was published. It is the largest continental atlas in China today. Moreover, soon after the Chinese expedition team surveyed it, China produced the first Changcheng Station topographical map – a symbol showing that China has joined the research and exploration effort in Antarctica.

Provinces and autonomous regions have published their own provincial (regional) atlases and general geographical maps at medium scales. These cartographic works reflect the basic situation of the political, economic and natural conditions in these provinces (regions). Cartographic developments are also taking place in synthetic and thematic mapping. For example, in 1985 China published an excellent synthetic thematic atlas, *The Physical Geography Atlas of Shanxi Province*.

In recent years, compilation and publication of thematic maps in China have been especially active in terms both of high volume and speed of production. For example, geological map series at 1:500,000 and 1:1 million scales of several provinces (regions) have been produced. *China's Palaeogeographic Atlas*, *China's Lithospherical Dynamics Atlas*, *Atlas of the Dryness and Wetness during the Past 500 Years*, and *China's Historical Seismological Atlas* (covering Ming Dynasty times) have also been published and these works reflect the research achievements in the related natural science areas over the past 20 or more years.

The development of agricultural mapping is even more rapid. Along with the nation-wide general investigation of agricultural soil and of agricultural planning, China has produced map series of agricultural soils at the scale of 1:500,000 and the *Soil Atlas of the People's Republic of China* has also been published. Some provinces (regions) and many counties have published their own atlases of agricultural regional planning or map series on the same theme.

Many environmental maps have been published, including urban environmental atlases and water protection research atlases. The first mapping work to summarise China's comprehensive natural resources and their utilisation and protection will be the *Nature Protection Atlas of the People's Republic of China*. Compilation is also planned of an *Atlas of the Yellow River Basin* and an *Atlas of the Yangtze River Basin* to show the natural relief and the rerouting and harnessing of China's two greatest rivers.

Population mapping is a new development. China has already published the *Atlas of Population of the People's Republic of China* based on the third national Population Census. The English edition of this atlas will be distributed overseas. In addition, the *Atlas of the Aged Population of the People's Republic of China* and the *Atlas of the Population of the*

Jiangsu Province have been published. Many of the maps in these atlases are drawn using Computer-Assisted Cartography (CAC).

With the rapid progress of China's tourist trade, many cities have now published their own tourist and transportation maps. Some are published in foreign languages or in a combination of Chinese and foreign languages. The *Atlas for Automobile Drivers* and the *Atlas of China Highway Transportation* have been well received by the users and the sales volume has reached nearly a million copies.

Urban mapping is also developing. Shanghai, China's largest city, has published the *Atlas of Shanghai*. Shenzhen, one of the first cities opened to the outside world, has also published its own atlas, the *Atlas of Natural and Economic Resources of Shenzhen*. These two comprehensive atlases of medium size are of high quality, both in their content and their printing techniques, and they symbolise the progress of China's urban mapping. Other cities are also planning to compile city atlases of their own and will develop further the concept and form of city planning atlases.

The great variety and numbers of China's maps and atlases published and distributed in recent years indicate that, when a country's economy is developing and its society is open, its cartographic undertaking inevitably gains in many respects. One indication of this is the need to provide extra printing and publication capacity: to meet the growing needs for both map making and publication, seven provinces have established their own cartographic publishing houses to add to the China Cartographic Publishing House, which is the largest publisher in the nation.

Cartographic Education

The development of cartography is largely dependent on the training of cartographers. The Wuhan Technical University of Surveying and Mapping (WTUSM), a higher institution specialising in the education of surveying and mapping experts, has been established since 1956. It contains a complete set of departments involved with surveying and mapping. Among them is the Department of Cartography. Over the past 30 years, a large number of cartographic specialists have been sent from here to different organisations in China. Many of the graduates have become the 'back-bone' of cartographic production, education and scientific research. In addition, there are three universities (colleges) other than WTUSM which train geographer-cartographers and geologist-cartographers at the university level.

Since the late nineteen seventies, post-graduate courses at the MSc level for majors in cartography have been set up and, in the mid-nineteen eighties, post-graduate courses in this field for a doctoral degree began to be offered. At the lower levels, there are cartographic branches at middle vocational schools. After two to three years' training, the successful students can take a job as cartographic technicians. The training of such specialised *cadres* with different levels of knowledge appropriate to different tasks can guarantee success in meeting the various national cartographic needs. The most important matter is the rational planning of the number of personnel trained at different levels.

In the numerous organisations in various provincial (regional) bureaux for surveying and mapping, there are many young workers and apprentices. These young people earnestly demand to learn new knowledge and to improve their specialist qualifications but, in China, it is impossible to establish sufficient schools to enroll all of them. Therefore, a new and appropriate educational form was created in China. In some provinces (regions) and functioning under the provincial (regional) bureaux, professional schools for surveying and mapping at middle levels of sophistication were established. Such schools can meet the demand for training from the younger generation, but many young cartographic *cadres* have been trained locally to meet local needs. Another successful mode of training is the cartographic correspondence course at WTUSM. Through part-time study for three to six years, young people can get a BSc degree after a successful completion of an examination.

In order to enable working cartographers to up-date their knowledge and to learn advanced cartographic technology, seminars on different themes have been conducted by the universities at convenient times. Seminars on computer-assisted cartography, thematic mapping, map reproduction and map printing have been warmly welcomed by young and middle-aged cartographers. Some of the seminars are internationally organised: under the cooperation with ICA, two seminars have been held in China since 1981. They were a 'Seminar on CAC' (1981) and a 'Seminar on Advanced Cartographic Education and Training' (SACET, 1986). Workshops headed by Professor F. Ormeling (President of ICA from 1980–1984) and Dr J. Morrison (President of ICA from 1984–1987) included many cartographic professors and experts from overseas and from China. Lectures and exhibitions were given to more than a hundred Chinese colleagues. Such activities strengthen academic exchanges between the Chinese and foreign cartographers. In addition to these lectures by famous cartographers, other overseas professors are invited to visit and lecture in some of the higher institutions each year. This makes it possible for Chinese colleagues to hear of international trends and the latest achievements of modern cartographic science and technology in good time.

Scientific Research in Cartography

Besides WTUSM, under the NBSM and the Provincial Bureaux for Surveys and Mapping (the PBSMs) there are research institutes for surveying and mapping, in each of which a cartographic research department has been established; moreover, some of China's geographic and/ or geological research institutes also have research departments of this kind. In total, therefore, there is quite a large research force in the entire country. These units not only deal with conventional cartographic research, compiling and producing regional thematic maps, but also pay more and more attention to computer-assisted cartography. In those units where economic conditions are better, CAC research is being turned to production use. Preliminary achievements using CAC methods have been obtained in making thematic maps and atlases.

In theoretical cartography, research on the nature of the map and map structure and pattern of cartographic communication are being dealt with in several different ways. Experimentally-based research in cartographic pattern recognition is being carried out in depth. Some cartographers have also put forward the concept of comparative theoretical cartography. They hold that the comparative method should be a basic method in the science of cartography, map reproduction and research activities. It is very likely that the method would be quite significant for map-making and map use if more investigation is made on its expression in cartography, its characteristics and its rules of application.

Research in CAC has reached a new level of sophistication. In May 1987, the Second China CAC Conference (CCACC-II) was held. Reports from the conference indicate that CAC in China has reached the stage of research on some essential cartographic problems, a significant advance from the lower level (such as simple automatic drafting) of several years ago. The essential problems are how to expand the functions of cartographic databases; the creation of new types of databases (such as relational and/or hypergraph types); the application of graph theory in cartographic databases; and the application of fuzzy mathematics and expert systems in cartographic generalisation, pattern recognition and various kinds of information systems based on cartographic and attribute databases.

In an age where the capabilities of the micro-computer are rapidly developing and becoming appreciated – especially in a country like China – the study and application of such micro-computers in map-making is of great significance. Several cartographic software packages of practical utility have been developed and some of them are being used in production.

It is becoming more and more popular to compile and up-date topographic and thematic maps using data from remote sensing and aerial photographs. There have been several major achievements in map production produced from remote sensing imagery. Two examples can be given here, the *Atlas of the Environmental Quality of Tianjin* and the *Atlas of the Geothermal Resources in Tengchong Area, Yunnan Province*. After five years' work, China's first remote sensing image processing system for map-making was established in 1987. This system can be used in research involving area measurements of land, forest and water resources, in the dynamic monitoring of land use and afforested area, in engineering surveying and designing, in producing and up-dating thematic maps, and so on.

Not all cartographic research involves computers. For instance, China has a long history in cartography: the maps recorded in Chinese historical literature can be traced back thousands of years. There are many large lithographic maps with a history of several hundred or a thousand years. Thus historical cartography is an important research area. In the last few decades, some specialists have written monographs on cartographic history through intensive study and collection of information. Recently, a new chronology has been established by some scholars which divides the phases of Chinese cartographic history according to those of general science. For example, the history of cartography's development can be divided into four periods. These are the pre-science period, the quasi-conventional science period, the conventional science period and the neo-conventional science period. Each period of cartographic history has its own corresponding characteristics. This process has been thought of as scientific cognition and classification. At present, a special group has been formed in the Research Institute of Surveying and Mapping under NBSM to conduct research on the design and compilation of the *Atlas of Ancient Chinese Maps*.

To ensure high quality, map production needs a standardised procedure. China has set up the Research Institute of Surveying and Mapping Standardisation to study and formulate the operating rules for surveying and mapping. Based on extensive investigation, this institute has worked out standard forms and specifications for topographic mapping and is still drafting various standards for map production.

In addition to the existing *Acta Geodetica et Cartographica Sinica* and *Bulletin of Surveying and Mapping*, the China Cartographic Publishing House and the Wuhan Technical University of Surveying and Mapping have published the first Chinese cartographic professional quarterly, *Cartography*, since 1986. It has been welcomed by a broad range of Chinese cartographers. China has also compiled a *Reference Handbook of Cartography*, published in 1987. This book acts as a small cartographic encyclopaedia for Chinese cartographers.

PROSPECTS FOR DEVELOPMENT

The earth, which maps try to reflect, is an integrating entity. Map quality is dependent on the thoroughness of investigation of all parts of the earth by mankind. Therefore, cartography, driven by its research objectives, must itself be an 'open subject'. China has taken an active part in international cartographic activities since we became a member state of ICA at the sixth General Assembly of ICA in 1980. Here, we would like to recall a significant historical event. The then ICA President, Professor Ormeling, and the Secretary-Treasurer, Mr. Hedbom, visited China for the first time in 1978 and had talks with the leaders of the Chinese Society of Geodesy, Photogrammetry and Cartography about China becoming a member of the ICA. When recollecting this event, one can, with respect, appreciate the foresight and sagacity of Professor Ormeling and his colleagues.

In order to develop a cartographic undertaking, a country has to define a strategy appropriate to the national conditions. China is a developing country with a vast territory and tremendous manpower, though its economic base is not yet so substantial and its standard of science and technology is at a lower level than those of the advanced countries of the world. We have to give full play to the great potential of the existing forms of map production so as to satisfy the requirements for cartographic output generated by the needs for national construction. Advanced large-scale computerised mapping systems, as a means of research, have been used in very few organisations and in institutions of higher learning and some gratifying achievements have resulted. However, it will be impossible for such methods to become a means of production in various map producing agencies in the near future.

Nonetheless, we should be aware that we are living in an era of cartographic transformation. While the traditional cartographic theory, cartographic technique and conventional cartographic products will still play their roles for quite a while to come, the electronic techniques and new types of cartographic products other than conventional maps will gradually contribute to all facets of man's activities. Therefore, researchers in cartography should be in contact with and explore these new fields while, in cartographic education, emphasis should be placed on connections between the old and the new. That is, students should have a good command of conventional cartographic theory, techniques and production as well as basic know-how and practice of advanced technology. If this is achieved, they can be competent at their present jobs and at the same time have the potential for further development.

The comparative method is useful in developing China's cartography because, with this method, we can take measures of focal points in accordance with China's links. For example, the traditional production style in map printing is a weak link technologically and it has affected map quality directly. Therefore, research and education in printing techniques have been especially set up in the WTUSM higher education institution, the aim of which is to enrich the technical force intellectually. We think this is the correct approach to take.

Since pursuing open external policies, China has achieved remarkable developments in economic matters and in the cause of science and technology. These have also been achieved in the field of cartography. One can see the results of increasing international contacts and many positive achievements in China's cartographic production, education and scientific research. Chinese cartographers are willing to promote international academic exchanges still further.

II

THEMATIC ISSUES IN CARTOGRAPHY

CARTOGRAPHY AS AN ART

Arthur H. Robinson

INTRODUCTION

To consider the proposition that 'Cartography is an Art' one must first define terms. 'Cartography' is here assumed to be map-making in the broadest sense, namely the preparation of a graphic representation of the milieu (Robinson and Petchenik, 1976, 1–17), including portrayals in three dimensions.

The term 'art' is another matter. There are innumerable activities with which 'the art of . . .' may be coupled, ranging from book design and wine-making to manuscript illumination and the writing of fiction. The only elements common to all things artistic seem to be that it involves a creative effort requiring some essentially intuitive judgment and that it invokes a subjective reaction in a knowledgeable observer or user. I make no attempt to define 'craftsmanship', 'skill', 'aesthetics', or 'beauty', all of which seem to be involved in the general concept of art.

If one looks up 'Cartography' in a dictionary, the definition is likely to be brief and similar to 'the art and science of making maps', as is also the ICA definition (ICA 1975, p. 1). The unanimity of the lay public and cartographers themselves suggests that there is no question that there is an artistic component to map-making. To examine this feature of the field, this paper will consider the following questions: What kind of art is cartography? What kinds of maps are likely to be more artistic? What design elements have the greatest potential for artistic treatment? and How might one interpret a map as a work of art?

WHAT KIND OF ART IS CARTOGRAPHY?

To consider the philosophy of art is to enter a murky semantic world. Nevertheless, there seem to be some basic concepts that will help to locate maps in the broad realm of art. The most fundamental is that a work of art is a man-made artifact distinguished from something natural. Another distinction, upon which there seems to be general agreement, is that any man-made thing responded to aesthetically is a work of art. That category of things is usually divided into 'fine art' versus 'useful art'. In the visual realm, this divides non-purposeful art objects (such as painting, abstract drawing and sculpture), that are appreciated for their own sake from things that are enjoyed as a means to something else. Useful art thus has both an aesthetic and a utilitarian dimension. Since all but fanciful maps are representational, being made to portray some aspect of reality, it seems logical to call cartography a useful art. Even if we prefer to be more restrictive and limit the meaning of the term 'art' only to those objects to which the response is solely aesthetic, one can argue convincingly, as does Keates, '. . . that the map, like other human artifacts, contains artistic possibilities' (Keates, 1984).

A reasonable parallel is architecture, the art and science of building, which attempts to produce structures that combine materials and space in useful ways that also evoke aesthetic reactions. But cartography has an extremely important additional role in that a map defines reality. Except from space, one cannot 'see' any large region in its entirety other than by mapping it. For example, neither Leif Ericsson nor Christopher Columbus 'discovered' North America; they only came across some land, and even Columbus was convinced until he died that it was something else. It remained for others actually to discover the continent by mapping it.

Cartography, like painting, is a visual art as opposed to literary or auditory art and is spatial in the sense that it is concerned with the arrangements of things in at least two dimensions, as opposed to temporal art such as music or literature. The distinction becomes cloudy when a single map tries to portray the changing features of a battle or when varying conditions are incorporated in a weather map on television.

In an important sense, all cartography is an art in that it is representational and that operation always involves some degree of abstraction. Geographical reality is infinitely complex and its complete depiction is quite impossible; elements must be left out and intricacies modified as a consequence of the fundamental requirement of information reduction. All maps, therefore, are abstractions and the decisions involved in the process are artistic in the sense that many of them must be made subjectively by the cartographer. The cartographer must integrate a great variety of elements, such as the map-maker's interests and understanding, the capabilities of the audience, the objective, the constraints imposed by the media, the costs of production, and the availability of data. The development of generalising computer programs involves these kinds of matters in the decisions, and the 'goal state' in a map design algorithm can be no more objective (Wilkinson, 1987). Many of the influences operating on the cartographer are not quantifiable, so that the result can only be ascribed to non-objective 'judgment' – a clearly artistic characteristic. Even the most rigorously-prepared, accurate, large-scale topographic map is an artistic creation. If a series is designed by a committee it makes no difference; a collective, subjective decision is still subjective.

WHAT KINDS OF MAPS ARE LIKELY TO BE MOST ARTISTIC?

As a broad generalisation, there may well be an inverse correlation between map scale and the artistic quality of maps. The large-scale category of plans that encompasses the detailed recording of cities and rural areas, including the modern computer equivalent of separate 'overlay' categories of geographical information, has small opportunity to appeal to the aesthetic. Except for the pleasure of satisfaction and appreciation to be derived from the recognition of more up-to-date accuracy and the technical freedom to interact (through Geographical Information Systems or GIS), such plans have little to offer as a graphic composition.

The somewhat smaller-scale, standard topographical map has more design possibilities. Like the oblique view of a landscape painting, the vertical view balancing the expression of natural and cultural features can evoke similar feelings in an appreciative viewer. To capture the essence of a landscape requires that the components be blended graphically so as to have an iconic quality, a unique sense of place and character. This aspect of topographic mapping is rather like portrait painting in that the objective is to produce an image blending feature and expression that conveys the essence of a personality. There is a feeling that mapping, with its planimetric constraints, is consequently too rigidly controlled to allow a parallel objective. This is hardly a reasonable view since a portrait painter must (usually, at least) put eyes, nose, ears and other features in their correct relative positions. A more significant restriction is the bureaucratic policy that symbolisation must remain standard throughout a range of such 'landscapes', resulting in a kind of cartographic homogenisation.

The older is a flat map or a globe map, the more likely it is to be called an art object. The medieval *mappaemundi*, the ornate portolan charts, the Ptolemaic maps and their later 'modern' additions are classed as artistic partly because they are old, partly because they seem to the modern eye (though not at the time) to be somewhat imaginative and partly because they are embellished with interesting decorative elements. Although in their culture they were serious historic-geographic documents with significant didactic purposes, they were also then considered works of art (Kish, 1986; Woodward, 1987b).

If that group of maps is evaluated by modern standards for their intrinsic artistic quality, a considerable number would probably be set aside. Because an 'art object' in the broad sense comprehends so many kinds of reactions, influences and evaluations, this paper is limited to maps and map design, disregarding such factors as historical significance, decorative elements, age, and the place of such maps in their contemporary culture. That is not meant to imply that such matters are not relevant to the basic topic of 'cartography as an art' but is rather a response to the inability to deal comprehensively with such a large topic in a short paper. The literature is extensive (see especially Woodward (ed.), 1987).

Some of the older maps do not fall in the above category, such as those produced by such artists as Leonardo da Vinci, Albrecht Durer and others in the Renaissance (Woodward, 1987a, pp. 1–9). As the topographical map developed, it became increasingly a military and engineering tool but in many cases it retained some of its iconic quality, especially when it included realistic expressions of terrain (Brown, 1949; Harvey, 1980; Tooley, 1978). Some of the 19th century European series are outstanding examples and are often described as 'beautiful'. The modern shaded relief editions of the US Geological Survey quadrangles are similar and, although somewhat mechanical looking, contrast significantly in their artistic quality with the less expressive orthophoto maps.

The mapping of the land form is clearly an artistic problem. Its sensitive portrayal requires an understanding of the response of diminutive man to the three-dimensional world in which we live. Granted that the greater the relief and thus the 'grander' the landscape, the more opportunities there are for expression, nevertheless man is so small in relation to even modest ups and downs of the terrain that capturing a feeling for landscape is an artistic challenge. Problems of generalisation required by scale, such as the elimination of detail and the sharpening of ridges or the arrangement of light and shadow, suggest that any entirely mechanical 'scientific' approach is not likely to produce truly artistic results in a two-dimensional map. The sculpturing of a three-dimensional model is similar in that it, too, requires attention to the sorts of generalisations required by the reduction in scale and the necessity for the artistic selection of the necessary vertical exaggeration to fit man's perception – perhaps even to be varied from place to place.

The observation that the smaller the scale, the more likely are maps to be artistic, needs qualification in the small-scale category. Small-scale reference maps may be appreciated for their skilful selection, generalisation, and integration of various kinds of geographical data. No matter how general maps characterise shapes, coastlines, rivers, transportation facilities, densities of populated areas, or whatever, such maps are basically geographical dictionaries. A dictionary can be greatly admired and appreciated for its accuracy, completeness, descriptive quality, typographical clarity, and so on, but it is not likely to be classed as a work of art. The thematic map, at the other end of the continuum of small-scale maps, has much more potential for being a work of art.

The (usually) small-scale thematic map that concentrates primarily on the portrayal of one geographical distribution (including the land form) or on some particular relationship can be characterised as a kind of geographical essay, an interpretational statement or assertion about something. The selection of techniques of scale, of tones and colours, the manipulation of the data, the allocation of emphasis, the development of figure-ground relationships, all are involved in the careful structuring of such a graphic essay. Thematic mapping provides a great many opportunities for distinctive graphic organisation. Given one set of data, such as from a census of population, literally scores of significantly different maps could be prepared. Many, although perhaps useful, would no doubt be dull and pedestrian but some could be classed as artistic.

With respect to both large- and small-scale maps, the steady increase in technical dependence that began to speed up in the 19th century, and has been accelerating ever since, has steadily affected the artistic component of cartography. The breakdown of the cartographic process into distinct elements and the control of these by consistent systems tend to displace the creative component. To be sure, interactive capabilities can be built into systems, but they have limited application. It has been said that what can be explained to a computer is science and that everything else is art. We are explaining more and more to the computer with a consequent increase in consistency and efficiency, but it is being accompanied by a decrease in the creative element (Woodward (ed.), 1987, p. 8).

WHAT DESIGN ELEMENTS HAVE THE GREATEST POTENTIAL FOR ARTISTIC TREATMENT?

A painter arranges a variety of graphic elements to complete a composition. These include colour, texture, shadow and highlights. In addition, various elements of structure are manipulated such as figure–ground, perspective and balance and all these are integrated to fit a total design that satisfies the artist. The artist works within the limits of the constraints imposed by the subject matter. Even the unconventional combinations of the surrealist require that things be recognisable but, in general, reality is required if there is subject matter. (The juxtapositions of arbitrary shapes, colours, textures, and so on of abstract painting are exceptional.)

Similarly, cartographers produce a composition by subjectively combining elements such as projections, colours, typography, line weights, symbolism, layouts and figure–ground relationships. If the innumerable decisions are made in truly artistic fashion, each element is considered simultaneously in relation to all the others to produce a total design that does not appear 'wrong' in any respect – considering, of course, the geographical constraints. Obviously, there is no one 'perfect' design, since a different choice of one graphic element changes the relationship amongst them all. The greatest freedom in this aspect of cartography as an art is provided by smaller-scale general and thematic mapping. In larger-scale general mapping, the objective is often to give equal visual weight to each category of data within the limits prescribed by tradition and convention; sometimes particular elements are given graphic priority, such as the landform or the transportation net. In much thematic mapping, the chosen topic is displayed with priority assigned to some aspect, such as the total structure of the distribution, its internal variations or its positional relationship to another distribution. In general, the smaller the scale, the more 'artistic' are the possibilities, culminating in world maps where a total conception of the earth is involved.

Map Projections
It is with world maps that considerations of map projections become most important. In the history of cartography it is claimed that the

Ptolemaic grid system became a powerful 'talismanic' symbol in that:

'. . . the cartographic grid of the Renaissance was believed to exude moral power, as expressing nothing less than the will of the Almighty to bring all human beings to the worship of Christ under European cultural domination.' (Edgerton, 1987, p. 12)

The 15th – and especially the 16th – century saw the innovation of numerous 'cordiform' (heart-shaped) projections that suggested both the 'living' character of the earth as well as its rotundity (Kish, 1965). Since then, there has been a continuous growth in the number and variety of map projections. Many of these, particularly conformal and equal-area varieties, have been devised to satisfy particular utilitarian requirements but a significant number have been aimed at 'artistic' applications. The potential role of projections in map design has become increasingly appreciated but is even yet often overlooked (Hsu, 1981).

The creative, perspective view of the earth, made popular by Richard Edes Harrison in the 1930s and continuing during World War II, is an example of the artistic employment of map projections at its best (for example, see Harrison, 1944). As the 'air age' changed into the 'space age', so the view of the earth as a spherical object has gained favour compared to simply a flattened display of it. Seeing the earth from a single viewpoint is often much more revealing – and artistic – than when our contemplation is restricted to the mechanical, orthogonal plan view. For maps of larger scale, an artistic objective might well lessen our insistence on a strict geometric framework for maps and make room for the greater use of mental constructs of social, cultural, and economic space (Watson, 1979; Robinson, 1979). Such maps might well be considered the cartographic equivalent of 'mild' surrealistic art.

Colour
Throughout the development of cartography, colour and the other primary graphic elements (size, shape, spacing, orientation and location) have been the media by which symbols were made distinctive. Colour components – primarily hue but, to a lesser degree, value – were the most important in the manuscript era, in the hand-colouring of printed maps, as well as in the recent period of printed colour. Earlier uses of colour were mostly for coding different elements of reality but:

'Overall in modern cartography there has been increased exploitation of graphic concepts of artistic design, scientific observations about depth perception and the figure–ground relationships of Gestalt psychology.' (Ehrensvard, 1987, p. 144)

The present freedom to obtain almost any colour qualities in printing and in electronic displays suggests that colour as an artistic element in cartography will become increasingly important.

Some of the more impressive artistic employments of colour have been in the representation of the land form, especially of mountainous regions. In this, the Swiss cartographers have reached impressive levels in the use of subtle hues of violet, yellow, blue and green to suggest the grandeur of mountains in school maps and the *Landeskarte der Schweiz* (Imhof, 1965). At any small scale, the portrayal of the land form by coloured hill shading is almost entirely an artistic exercise, since anything approaching realism (except as to general location) is quite impossible because of the requirements of generalisation and exaggeration.

Typography and Lettering
The consensus is that names are generally necessary on maps for a variety of functions, such as labelling places, indicating extent and qualitative identification. The typography is often the focus of evaluation, both functionally in terms of its size or discriminability and aesthetically in terms of style. Until the 19th century, the names on maps in western cartography were largely either lettered by hand (earlier) or engraved by hand (later) but, since the mid–19th century, they have been mostly typographic (Woodward, 1987a, pp. 175–6). The later hand lettering and engraving tended to follow the typographic designs of the period, reinforced by the technical necessity of actually incorporating type in various ways on printed maps. Only since the first quarter of this century have type faces been seriously designed with their use on maps as a significant element in their creation; the first was that by the National Geographic Society for its well-known and distinctive-looking series of maps. (The distinctive appearance is largely a result of the type styles.) The incentive was not dissatisfaction with available styles of type or hand lettering styles, but was based upon the slowness of hand lettering, the few practitioners of the art and (especially) the fact that available type styles could not withstand the necessary enlargement and reduction in photo-composition for the modern stick-up lettering process.

The arts of calligraphy and typography clearly have a long association with cartography. As Woodward observes:

> 'Lettering has been a prominent and ubiquitous element on maps. Its style is so sensitive to regional and historical differences in taste that it can be used, along with other cartographic elements such as symbolism, colour and iconography, as a period guide to identifying maps.' (Woodward, 1987a, p. 212)

There are obviously other design aspects of cartography that enter into the total artistic content of this 'useful art'. These include symbolisation, comparative line weights, balance of structural elements within the map frame and the employment of figure–ground relationships. All these must be combined with a projection, use of colour and lettering to produce a total composition. All are important but, when one surveys the past, the greatest number of innovations and the largest controversies have concerned the three aspects of projections, colour and lettering. More often than not the comments and debate have focussed not on their utilitarian aspects but on aesthetic reactions. Terms such as 'mis-shapen' (continents), 'garish' (colours), 'ugly' (lettering) and similar characterisations are countered with opposite reactions such as 'appropriate', 'pleasing' and 'delicate'. Clearly cartographers themselves feel that projections, colour and lettering are important artistic aspects of cartography.

HOW MIGHT ONE INTERPRET A MAP AS A WORK OF ART?

There are a number of factors one should consider when evaluating a map as a work of art. These include other works by the cartographer, works of the same kind by other cartographers, relevant facts associated with the production of the map, and a study of the period when the map was produced.

The first of these factors – other works by the same cartographer – is significant for it allows us to assess to what degree the cartographer employed the same or different line weights and character, lettering styles and positioning, colours and other elements of contrast and harmony. For example, a comparison of August Petermann's early production while in London in the mid–19th century with his later productions in a series in *Petermanns Geographische Mitteilungen* reveals his growth as a cartographic artist (Robinson, 1982). Similarly, Charles Joseph Mindar's long series of flow maps shows steady development from his first relatively unsophisticted production (Robinson, 1967; 1982). A good many general – and especially thematic – maps are single or only occasional productions by a cartographer. In the more recent history of cartography, there are relatively few career cartographers – such as Petermann – whose production over a considerable period allows us to reflect on the maturation of their talents as artists.

A study of similar maps by other cartographers at the same period allows one to develop an appreciation of the relative capabilities of the cartographer. Certainly the quality of Guerry's 1864 maps of moral statistics can be better appreciated when compared with those by Mayhew and Fletcher (Robinson, 1982, pp. 156–70).

Of major significance in the appreciation of a map as a work of art is an understanding of the media available to the cartographer. From at least the time of the Middle Ages, manuscript maps on vellum and (later) paper were produced under relatively few constraints. Many inks, paints, brushes and quills were available. Other surfaces – such as birch bark, silk, horn and plaster on walls – considerably decreased the options. Only in more recent times did the development of materials and equipment, such as grained paper and the air brush, again enlarge the opportunities for manuscript production. The advent of computer-produced maps on monitors and TV screens has made available enormously enlarged capabilities for colours, lines and – especially – animation. The common 'manuscript' map of today is the television weather map. In the United States at least, these are usually locally produced, so the mobile population has the opportunity to compare similar maps by a variety of cartographers.

Although the television weather map is literally reproduced by the tens of thousands – or even millions – on TV screens, the reproduction is simultaneous. The successive reproduction of identical copies, beginning in the 15th century, placed a considerable constraint on the cartographer. To judge and appreciate such non-manuscript maps, one must be familiar with the capabilities and limitations of whatever process was utilised to make the copies (Woodward (ed.), 1975).

An interpretation of a work of art is aided by a knowledge of the period when it was produced, encompassing the social, scientific and economic contexts as well as the availability of data. The concern of mariners for information about sea and air currents, tides and compass variation – coupled with a knowledge of the growth of understanding and the attempts to develop basic theory – greatly aids appreciation. The degree of analysis and theorising involved in some of the early maps, such as those by Halley, should not be compared with present productions of the same *genre*. A knowledge of the social upheavals accompanying the industrialisation of western Europe in the 19th century aids in appreciating the class of moral statistics and sanitary maps of the period.

Biographical information about the cartographer helps one to understand the motivation and intentions of the map maker. The training, experience and associations of a cartographer are all relevant. For example, the statistical expertise of Adolphe Quetelet helps one to appreciate why he introduced continuous tone shading of 'the darker, the more' as a substitute for the choropleth technique, with its sharp visual divisions, employed earlier by Dupin (Robinson, 1982, p. 160). In that instance, we have the stated intentions of Quetelet. Unfortunately that is the exception for, as has been observed:

> 'maps. . . are inarticulate and their silence seems to have affected their makers. It is as if their expertness in the graphic expression of facts was accompanied by an atrophy of verbal expression following its disuse.' (Davis, 1924, p. 194)

The question of intent on the part of artists in interpreting their works is controversial. On the one hand, it seems reasonable that, if a cartographer had a particular purpose in mind (ideally clearly stated) – whether it be to portray the character of the landscape by a general map or to focus on a particular aspect of a distribution by a thematic map – the series or map should be judged on that basis. On the other hand, when a work of art is made public, it ceases to be the 'property' of the maker but becomes public property – to be judged strictly on its artistic merit. Thus, if a topographical map like the *Carte de Cassini* does not include or badly characterises one important element of the landscape such as the land form, as a work of art it should be judged as deficient, even though its designer (Cassini de Thury) specifically stated he had no intention of including the land form (Dainville, 1959).

CONCLUSIONS

Maps *can* be works of art since there is nothing in the cartographic process that prevents it. To be sure, being a 'useful art' does add a utilitarian dimension to map-making that may tend to inhibit creative experimentation. On the other hand, the availability of modern electronic resources has immeasurably increased the opportunities for observing geographical space, manipulating it in various ways, and displaying the results graphically. Cartography tends to depend heavily on tradition but, even with our self-imposed inhibitions, there is plenty of opportunity to produce artistic maps.

REFERENCES

Brown, Lloyd, A. (1949), *The Story of Maps*, Boston: Little, Brown; reprinted New York: Dover, 1979.
Dainville, François de (1959), 'De la profondeur a l'altitude', Bibliothèque générale de l'École pratique des Hautes Études, VIe section, Paris: pp. 195–

213. Translation: 'From the Depths to the Heights', *Surveying and mapping*, 30, 1970, 389–403.

Davis, William Morris (1924), 'The Progress of Geography in the United States', *Annals Assoc. American Geographers*, 14, p. 194.

Edgerton, Samuel Y. Jr (1987), 'From Mental Matrix to Mappamundi to Christian Empire: The Heritage of Ptolemaic Cartography in the Renaissance', in Woodward (ed.) (1987), 10–50.

Ehrensvard, Ulla (1987), 'Colour in Cartography: A Historical Survey', in Woodward (ed.) (1987), 123–46.

Harley, J.B. and David Woodward (eds.) (1987), *The History of Cartography*, Vol. 1, *Cartography in Prehistoric, Ancient, and Medieval Europe and the Mediterranean*, Chicago: Chicago University Press.

Harrison, Richard Edes (1944), *Look at the World, The Fortune Atlas for World Strategy*, New York: Alfred A. Knopf.

Harvey, P.D.A. (1980), *The History of Topographical Maps: Symbols, Pictures and Surveys*, London: Thames and Hudson.

Hsu, Mei-Ling (1981), 'The Role of Projections in Modern Map Design', *Cartographica*, 18, 2, 151–86.

ICA (1973) *Multilingual Dictionary of Technical Terms in Cartography*, Commission II, ICA, Wiesbaden: Franz Steiner Verlag.

Imhof, Eduard (1965), *Kartographische Geländedarstellung*, Berlin: Walter de Gruyter & Co. English version, ed. by Harry J. Steward, *Cartographic Relief Representation*, 1982.

Keates, J.S. (1984), 'The Cartographic Art', *Cartographica*, 21, 32–43.

Kish, George (1986), 'Cartes, globes et arts decoratifs: une vue des géographes', *Acta Geographica*, 3rd series, 66, 65–81.

Robinson, Arthur H. (1967), 'The Thematic Maps of Charles Joseph Minard', *Imago Mundi*, 21, 95–108.

Robinson, Arthur H. and Barbara Petchenik (1976), *The Nature of Maps*, Chicago: University of Chicago Press.

Robinson, Arthur H. (1979), 'The Image and the Map', *Congress Proceedings*, 23rd International Geographical Congress (Montreal), Ottawa, pp. 50–61.

Robinson, Arthur H. (1982), *Early Thematic Mapping in the History of Cartography*, Chicago: University of Chicago Press.

Tooley, R.V. (1978), *Maps and Map Makers*, 6th ed., London: B.T. Batsford.

Watson, J. Wreford (1979), 'Mental Image in Geography: its Identification and Representation', *Congress Proceedings*, 23rd International Geographical Congress (Montreal), Ottawa, pp. 38–50.

Wilkinson, G.C. (1987), 'The Search Problem in Automated Map Design', *The Cartographic Journal*, 24, 1, 53–56.

Woodward, David (ed.) (1975), *Five Centuries of Map Printing*, Chicago: University of Chicago Press.

Woodward, David (ed.) (1987), *Art and Cartography*, Chicago: University of Chicago Press.

Woodward, David (1987a), 'The Manuscript, Engraved, and Typographic Traditions in Map Lettering', in Woodward (ed.) (1987), pp. 174–212.

Woodward, David (1987b), 'Medieval Mappaemundi', in Harley and Woodward (eds.) (1987), pp. 286–370.

THE SCIENCE OF CARTOGRAPHY

Árpád Papp-Váry

THE DEMAND FOR CARTOGRAPHY

Social demands have always created new fields of science and have stimulated the development of existing ones. This well-known philosophical statement is as true for cartography as for any other science. The changes in the social demands behind the development stages of map-making can readily be traced, demand for maps having dramatically increased after World War II. The following factors in particular have accelerated developments in map making and map production.

(i) The rapid growth of home and international tourism.
(ii) The disintegration of colonial systems, the creation of development programmes in the third world countries and the establishment of extensive trade relations.
(iii) The increase of interest in remote areas of the Earth, facilitated by the developments in communications which have created a 'shrinking globe'.
(iv) The increasingly general use of methods in education involving practical work, varying in type with the age of pupils.
(v) The general spread of regional planning and development policies, which are motivated both by the process of urbanisation and the aim of solving or avoiding the social problems associated with industrial districts.
(vi) The assessment of environmental pollution, the organisation of conservation measures, the continuous survey of changes in the environment and the development of environmental assessment procedures.

These and other growing demands by society have resulted in the widespread use of map products and, as a consequence, led to a rapid increase in the number of mapping professionals and to the establishment of an institutional organisation of cartography.

THE EVOLUTION OF CARTOGRAPHY AS A SCIENCE

According to studies on the history of science, all sciences can generally be described as consisting of the following structural units (Pápay, 1983):

(i)   The subject matter of the science.
(ii) The manifestation of the subject and its aims.
(iii) Characteristic and particular methods of the science and its terminology.
(iv) Professionals involved in the science.
(v)   The institutional organisation.
(vi) Social relationships within the science.

These structural elements develop slowly and they continuously strengthen. A new science is almost imperceptibly born when all of these elements are present. We can now consider how the situation in cartography matches to each one of these requirements for a new science.

The subject of cartography – with certain simplifications – can be described as the study of geographic space or of the graphical manifestation of spatial phenomena. The object of cartography is to produce maps which are able to reflect reality as exactly as possible.

The special and basic methods of the science of map-making were established with the emergence of thematic cartography in the 19th century. After the evolution of relief representation and thematic mapping methods, the first cartographic textbooks were published in the last century, such as those by Toth in 1869 and Zondervan in 1898; they summarised the evolving knowledge of the profession and also sounded the birth of a new science. The earliest and best known synthesis of the science of cartography was written by Max Eckert in 1921. The first comprehensive books on thematic cartography were, however, only published in the 1960s.

The established terminology of the profession – at least in a limited number of the most widely used languages – was collected together in the Multilingual Dictionary of Technical Terms by the International Cartographic Association in 1973 (ICA 1973). Thus the dictionary served two purposes, providing a codification of the then commonly used terms as well as a translation mechanism.

The first independent, professional cartographers appeared in the last century. The cartographers of that century only became cartographers through their practical work; they originally were geographers, copperplate engravers, engineers and army officers. The formal education of cartographers and the first university departments of cartography only began to emerge in the 1950s and 1960s; the exception to this was the founding of the first department of cartography in the Soviet Union in 1936, an achievement far ahead of its time. The university developments in the 1950s were soon followed by the establishment of special secondary schools for the institutional training of technicians and drafters. At the present day, the educational structure has evolved to include even forms of post-graduate education in cartography – particularly in the field of remote sensing and automated map production.

This move to an independent existence for cartography and the evolution of the institutionalised education system were followed by the appearance and rapid increase in numbers of cartographic periodicals. Not only did the number of periodicals grow in the 1960s and 1970s, but they were also published more frequently. Often intimately associated with the publication of these professional journals was the creation and development of the national cartographic societies.

The foundation of national organisations and the publication of independent periodicals were also greatly helped by the establishment of the International Cartographic Association (ICA) in 1959/60. The regularly organised international meetings and the direct exchange of views therein have strengthened the consciousness of cartography and also accelerated its organisational independence in several countries. In addition to the international conferences of the ICA, the number of national and various international conferences and map exhibitions initiated by individuals has dramatically increased. Morover, our concern with information handling has grown: increasing emphasis is now laid on the importance of map collections and map catalogues as well as on the training and post-graduate education of map librarians.

By graphing the dates of the foundation of cartographic periodicals, university departments and cartographic societies with their numbers, a transition from the scattered data after the mid–1930s to a sharp increase between 1950 and 1975 becomes evident. This phase in the history of cartography was described by F.J. Ormeling in 1972 as 'turbulent, impetuous cartography'. After 1975, a slower but continuous growth can be observed. The three periods in the graph can be specified as the periods of preliminary growth, of exponential growth and that of levelling off. These intervals basically correspond to the phases in the growth curves of other subjects, as described in various works on the history of science.

To take another view on the emergence of a discipline, J.A. Wolter (1975) has argued that there are four distinct phases in the development of sciences.

(i)   The appearance or the germination of the science.
(ii) The development stage of the science.
(iii) The blossoming of the science. Characteristic of this phase is the appear-
      ance of contradictions which cannot be solved by traditional means
      derived from pre-existing science.
(iv) Its break-up into various sub-fields which form the nucleii of new fields
      of science.

According to Wolter's description, the state of cartography when he wrote corresponded to the rapidly rising part of the scientific growth curve – that is, cartography was then a developing science. F.J. Ormeling (1972) – and even later Gyula Pápay (1983) – have also written about cartography as a nascent science. Pápay reasoned that the general principles in cartography (generalisation, map use, thematic cartography) are only in their early stages as yet. K.A. Salichtchev, in his article prepared for the 1982 ICA conference in Warsaw, reminded us that the science of cartography was born in this century and hence was relatively juvenile in scientific terms.

To some degree, these differences merely reflect nuances perpetrated by use of different terminologies in different countries. The best solution is to regard cartography not as a nascent, but rather as a developing science. This new science became a separate and developed entity in the 1960s.

CARTOGRAPHY AND ART

The essential purpose of map-making has always been the creation of the most exact reflection of reality or the graphically true representation of space. The complex nature of reality can only be abstractly reflected. Map-makers have always been ambitious to create map products where the abstract representation is able to recall the image of the object – that is, the map readers are able to identify the graphic figures with their image of reality or, put another way, able to recognise reality 'in the image of reality' and from which they are able to extrapolate to the real world. To achieve these aims, cartographers in the past made their maps to high artistic standards. To increase the aesthetic effect of their products, the titles and legends were surrounded with artistic figures and the map frame was also artistically drawn.

The purpose of such artistic work was to help the recognition of reality on the basis of the use of the maps; at the same time, the attractive figures made the map readers interested in the map and its use. The basic purpose of the maps, however, was still to reflect reality as perfectly as was possible given the knowledge of the time. The scientific problems of the exact representation of the real world have always been the primary and determining factor in the process of map-making, while the artistic work has only been of secondary importance.

The romantic view of cartography – which considered cartography as a mixture of science and art ('cartography is the art and technique of map-making') has in fact never been accepted generally, though it has appeared in various publications. It is worth noting, however, that the first and generally accepted definition of cartography was given by the ICA, which stated that art is an integral component of cartography.

CARTOGRAPHY AS A FORMAL OR AS A COGNITIVE SCIENCE

According to the ICA definition, the scientific object of cartography is map-making; in this context, the term 'map' includes all types of maps, globes, three-dimensional models and sections. Thus, the ICA considers cartography a formal science; its aim is the cartographic representation of spatial information and the development of its methods. Cartography, like mathematics, develops in abstract forms and gives methodological assistance to natural and social sciences (Arnberger, 1966). Many professionals, then, consider cartography a formal science.

On the other side, several authors consider cartography a cognitive science. The object of cartography, according to them, is the cognition of reality by means of maps which graphically model the real world. The form (cartographic representation) cannot be separated from the content (represented reality). Development of representational methods is stimulated by the deeper cognition of reality. The maps help us answer not only the question of 'where?', but also 'how?', 'why?' and 'when?' (Salichtchev, 1980).

There have often been sharp disputes between those holding these two views. Although opinions, in the meantime, have somewhat changed and grown more sophisticated, the differences between them are still evident. A comprehensive survey of this problem would require a complete study. Nevertheless, on the basis of practical experience, the trend towards viewing cartography as a cognitive science seems to be more convincing.

The question perhaps will be answered by the on-going research on

Geographic Information Systems. The most prominent representative of the cognitive trend is K.A. Salichtchev; he stated that the future of cartography will be determined by the systematic mapping of the environment (socio-economic complexes and geosystems) and by the cartographic integration of localised information on nature, population, economy and culture. In systematic mapping, individual map elements (relief, vegetation, etc.) are only 'sub-systems'; thus the thematic maps will not be produced separately or individually but in close co-operation. This form of map production requires the theoretical evaluation of the internal and external relationships between elements and the working out of the hierarchical levels of individual elements.

This aim is effectively realised by the Geographic Information Systems (GIS). The task of cartographers working with the construction and management of such systems should not be restricted to the generation of graphic images of the real world. GIS is, in fact, a new device to obtain deeper knowledge about reality. Cartographers have, and must have, their role not only in the development of new devices, but also in the exploration of new areas of reality and of newly discovered spatial relationships. This requirement strengthens the cognitive function of cartography.

NEW CARTOGRAPHY

The system of cartographic knowledge is not yet established and the practical experience of cartography has not yet resulted in the formulation of laws (with the exception of the attempt by J. Pravda). Due to the lack of development of what may be called the metasphere (theoretical cartography, metacartography), several attempts have been made in the last two decades to work out the 'theoretical frames' of the science of cartography (metacartography, cartology, cartographic communication, cartographic psychology, cartographic semiotics, cartographic modelling and the language of maps).

Today, Computer Science is expanding rapidly; if cartography only develops step by step, it may well result in us falling behind and, finally, the problems in cartography may become solved in other fields of science. This danger demands a comprehensive renewal of the profession and the re-thinking of the theory of cartography on the basis of Computer Science. Some authors believe that cartography in the age of computers requires a substantially new basis and approach. This new science or profession – to distinguish it from the traditional cartography – is called New Cartography. However, a Pallas-Athene-like birth of cartography is rather unimaginable. Digital cartography can only develop on the existing basis and, for a time, new cartography must live side-by-side with the traditional or manual map-making.

The current task in cartography is to create theory for our science which can help to satisfy the emerging practical demands, which originates from the history of the science and which can be fully justified in practice. If we act immediately, our science may continue on the course of its development. If we do not . . . well, we should not think of that.

REFERENCES

Arnberger, E., *Handbuch der thematischen Kartographie*, Wien, 1966. Franz Deuticke.

Morrison, J.L., The science of cartography and its essential processes, *Cartographica*, 1977, 19, 58.

Morrison, J.L., Cartography milestones and its future, *Proceedings of Auto-Carto London*, 1986.

Ormeling, F.J., Turbulent cartography, *Geografische Tidschrift*, Delft, 1972, Uitgeverij Waltman, 1–19.

Ormeling, F.J., Einige Aspekte und Tendenzen der modernen Kartographie, *Kartographische Nachrichten*, Gütersloh, 1978, 90–5.

Pápay, Gy., A Kartográfia-történet korsza-kolásának módszertani kérdései (Methodological problems in periodising the history of cartography), *Geodézia és Kartográfia*, 1983/5, Budapest, 344–8.

Papp-Váry, A., Ein Handbuch für Kartographie aus dem voringen Jahrhundert: ein Werk von Ágoston Tóth, *The International Yearbook of Cartography*, XXIII, Bonn-Bad Godesberg, 1983. Kirschbaum Verlag, 105–19.

Pravda, J., Kotázke kategorii a zakonov v kartografii (Categories and laws in cartography) – *Geodeticky a Kartograficky Obzor Praha*, 1983/12, 307–313 pp.

Salichtchev, K.A., *Idei i teoreticheskie problemi kartografii*, 80-h godov Kartografia 10, Moscow, 1982, VINITI, 156 pp.

Taylor, D.R.F., The educational challenges of a new cartography, *Cartographica*, 22, 4, 1985, 19–37.

Wolter, J.A., Cartography – an emerging discipline, *The Canadian Cartographer*, Toronto, 1975/2, 210–16.

HISTORICAL CARTOGRAPHY

Helen Wallis

THE ORIGINS AND EVOLUTION OF HISTORICAL CARTOGRAPHY

In a lecture delivered at the Smithsonian Institution, Washington D.C. in 1856, the German scholar Johann Georg Kohl remarked: 'If geography itself was neglected until our days, the history of geography must, of course, have been utterly unknown.' Descending the scale, he continued, 'If, as I have said, the history of geography has been utterly neglected, then I must add, that the most essential part of it, the history of geographical maps, has scarcely ever been thought of. . . . . This branch of geographical research remained a perfect blank until our days'.

The subject of 'historical cartography' or the 'history of cartography', as it is more usually known, was then beginning to make progress (Skelton 1972, p. 62). In 1839, the Vicomte de Santarem, the Portuguese *emigré* scholar in Paris, had coined the term 'cartographia' to describe the study of maps. The word was taken up quite quickly by geographers and was applied to the making of maps. Practitioners were soon calling themselves 'cartographers'. Both historical and contemporary cartography gained in prestige from the introduction of this scientific term to describe map-making.

The increasing interest in the history of cartography in the later years of the nineteenth century proceeded in parallel with the growth of geography. Historical cartography functioned in effect as the handmaiden to geography, as J.B. Harley has recently described it (Harley and Woodward 1987, p. 12). For various reasons, it was more closely associated with geography than with contemporary map-making. Historical studies in cartography needed the support and patronage of an established discipline. The subject had still to build up its own intellectual framework and was aided in this task by various geographical institutions founded earlier in the century. Two of the most outstanding of these were the Société de Géographie, established in Paris in 1821, and the Royal Geographical Society of London, formed in 1830.

At the same time, great libraries – such as the British Museum in London and the Bibliothèque Nationale in Paris – were setting up map departments to service and augment their rich collections. 'A public library is the safest port', Richard Gough had written in 1780, deploring the loss of so many topographical works in Great Britain (Gough 1780, p.xlvii). Thus a corpus of source material became available.

The publication of facsimile atlases in Paris by Santarem (from 1842) and, in a spirit of keen rivalry, by Edme François Jomard of the Bibliothèque Nationale (also from 1842) brought many early maps to the attention of scholars at large. Lithographic printing, moreover, permitted the reproduction of manuscript maps in colour. Adolf Erik Nordenskiöld, the explorer and collector, produced in his *Facsimile Atlas to the Early History of Cartography* (Stockholm, 1889) and his *Periplus* (Stockholm, 1897), two of the standard reference books for the history of cartography. The fact that facsimiles of these facsimile atlases have been published in recent years, Nordenskiöld's *Facsimile Atlas* by Dover Publications, New York, in 1973, and the *Atlas de Santarem, Facsimile of the First Edition 1849*, edited by A.H. Sijmons, by Rudolf Muller of Amsterdam in 1985, is testimony to their enduring value.

International Geographical Congresses provided a forum for the promotion of historical cartography. The first Congress, held at Antwerp in 1871, included an exhibition of rare early maps. The London Congress in 1896 gave the public an opportunity to see the maps and charts of the 16th century Dieppe School, important for the discovery of North America and also, arguably, of Australia. A Commission for the Reproduction of Early Maps was founded at the Geneva Congress of 1908. Its successor was the Commission on Early Maps, appointed by the Lisbon Congress of 1949 and dissolved at the London Congress in 1964. The first of its proposed four-volume catalogue, *Monumenta Cartographica Vetustioris Aevi*, appeared in 1964, entitled *Mappemondes* AD 1200–1500, edited by Marcel Destombes of Paris. A working group of the International Geographical Union, which succeeded the Commission, carried the project forward under the chairmanship of Professor George Kish of the University of Michigan. The second volume to be published, the fourth of the projected series, appeared in 1987 and is entitled *The*

Earliest Printed Maps, 1472–1500, by Tony Campbell of the British Library in London.

In 1935, the founding of the first scholarly journal on the history of cartography, *Imago Mundi*, was an indication that the subject had outgrown its subordinate status. *Imago Mundi* was launched as the inspiration of the Russian scholar Leo Bagrow, a man 'with fire in his belly', to quote R.A. Skelton (1972, pp. 99–100). Bagrow's intention was to make *Imago Mundi* 'an international centre of information'. Fifty years later, I have confidence in asserting, despite a contrary view recently voiced (Harley and Woodward 1987, p. 29; Harley 1987), that the succession of distinguished editors has fulfilled this aim. *Imago Mundi* is still the major scholarly journal in the history of cartography, although it has been joined by other publications meeting the needs of a more popular market.

After the Second World War, maps assumed a new importance – partly because of the increased publication and the greater use of maps by officials and partly because of the large consignments of cartographic materials which passed to libaries for preservation. In Europe and North America, many map collections had been unexploited hitherto because of lack of trained staff and proper catalogues. When the American historian Justin Winsor, librarian of Harvard University Library, had appointed a map curator in 1884, he greeted him with the words, 'Well, all I can do is to turn you loose, and let you flounder' (Wallis and Zögner, 1979, p. 107). Map librarians had to be self-educated, and library directors all too often regarded the appointment of a map librarian as a luxury which they could not afford. The British Museum – which was the exception to this general rule – had published a three-volume catalogue of its manuscript maps in 1844 and 1866 and its *Catalogue of Printed Maps, Plans and Charts* in two volumes in 1885. When a new edition of the *Catalogue of Printed Maps* appeared in 1967, the Museum could still claim to be the only national library to have published a complete (or relatively complete) map catalogue.

Within twenty years of the end of the Second World War, the history of cartography had become a flourishing subject in its own right. International developments both encouraged and reflected a new vigour and confidence. The setting up of the series of biennial international conferences from 1967 responded to the wishes of historians of cartography to maintain a sturdy independence of any parental organisation. The forerunner of the series was a one-day conference held in July 1964 as part of the 20th International Geographical Congress in London. The success of this meeting led a small group of us to plan a two-day conference in London in 1967, with R.A. Skelton (then recently retired as Superintendent of the Map Room of the British Museum) as Chairman and Professor E.M.J. Campbell as Secretary of the Organising Committee. The varied background of participants, some of whom pursued the history of cartography as a serious hobby, justified this policy. The conferences are run in association with *Imago Mundi*, and the host country always has a free rein in planning and organisation. Although *Imago Mundi* offers to publish selected papers, the host may arrange publication on its own account. In 1971, the Society for the History of Cartography was founded as a convenience for organisers and participants alike and included all subscribers to *Imago Mundi* (Fig. 6).

HISTORICAL CARTOGRAPHY AND THE ICA

Another major advance followed with the establishment of a working group in the history of cartography by the International Cartographic Association. The Association, founded in 1959, had accepted Stéphane de Brommer's definition of cartography, translated from the French as 'The art, science and technology of making maps, together with their study as scientific documents and works of art' (Ormeling 1987, p. 15). This definition would not be complete without 'giving recognition to the roots of its subject and taking official cognisance of the study of its heritage', as A.H. Robinson has remarked and as Fer Ormeling has reminded us in his history of the Association (Ormeling 1987, p. 57).

The Working Group took up as its first project the preparation of an historical glossary of cartographic innovations up to 1900. It published a pilot study in time for the ICA's International Conference at Moscow in 1976. The General Assembly at Moscow promoted the Working Group to the status of a Commission and, at Perth in 1984, it became one of the four Standing Commissions of the Association. In October 1987, at the next International Conference in Morelia, I – as Chairman – had the pleasure of presenting to each of the Presidents past and present a copy of the recently published volume *Cartographical Innovations. An International Handbook of Mapping Terms to 1900* (Robinson and Wallis 1987). It sets out to show how ideas developed, how processes and techniques began, when materials were first used, and how knowledge of innovations was diffused and transmitted. The project was organised

Fig. 6. Members of the Fifth International Conference on the History of Cartography, held at the National Maritime Museum, Greenwich in 1975. (By permission of the National Maritime Museum).

on an international basis, with contributions from 98 scholars in sixteen countries. Several countries established working groups to assist. The Acknowledgements pay warm tribute to the help and encouragement received from the Presidents, A.H. Robinson (1972–1976), Fer Ormeling (1976 to 1984) and Joel Morrison (1984 to 1987); they end with the hope that modern cartographers will find the work useful, since they too are producing the source material for future historians of cartography.

The terms of reference for the Commission for the period from 1987 to 1991 enlarge its role. The list of tasks and chairmen of the Working Groups are as follows:-

(i) Preparation of a source book for a biographical dictionary of cartographers, including reminiscences of 20th century cartographers.
-Helen Wallis (UK) and Monique Pelletier (France)

(ii) The teaching of the History of Cartography: to investigate the extent of university teaching of the history of cartography, course content, neglected topics, and the relationship to advanced research; and to indicate the role of libraries and archives in promoting the education of the public.
-R.I. Ruggles (Canada)

(iii) To build up a database of cartochronology up to 1930, for the dating of geographical elements of cartographical documents. This is designed to benefit historians of cartography, users of maps in general and the makers of historical maps and atlases.
-Tony Campbell (UK)

(iv) To prepare a source book of non-conventional cartographic systems (e.g. maps used in non-literate cultures).
-W. Scharfe (FRG)

The Commission has also run sessions on the history of cartography at the international conferences, one of the most successful being that at Perth, Western Australia in August 1984. This gained a high degree of interest from the media.

OTHER INTERNATIONAL AND NATIONAL DEVELOPMENTS

Thus, through the ICA, historical cartography has gained its proper place within the body politic of cartography. On a national basis, there have been some similar developments. The British Cartographic Society, for example, was founded in 1963 and set up a Map Curators' Group in 1966. This has encouraged an historical perspective in the work of the Society. The American Congress on Surveying and Mapping and the Australian Institute of Cartographers are similarly oriented, and the publications of all these (the *Cartographic Journal* (UK), *The American*

Cartographer and *Cartography* (Australia)) include papers on the history of cartography. In Canada, the periodical *Cartographica*, founded and edited by B.V. Gutsell of the University of Toronto Press and endorsed by the Canadian Cartographic Association, ranks among 'International Publications on Cartography', as stated on the title page. In the German Federal Republic, the Arbeitskreis Geschichte der Deutschen Gesellsch- alf für Kartographie runs historical colloquia and produces publications.

Other societies cater for more specialised interests. From 1952, the Coronelli Weltbund der Globusfreunde, founded by Robert Haardt in Vienna and now called the Coronelli Gesellschaft, has catered for the history of terrestrial and celestial globes. It has held international conferences and includes in its journal, *Der Globusfreund*, inventories of globes preserved in different countries. International reunions on the history of hydrography and nautical science are held every few years, alternatively in Portugal and Brazil. Papers are published by the Centro de Estudos de Cartografia Antiga, Lisbon and Coimbra. The Society for the History of Discoveries, based in the USA, runs annual conferences and its periodical, *Terra Incognitae*, includes studies on the history of cartography.

A further development since the Second World War has been the growth of private collecting and of a flourishing antiquarian map trade. New societies and new publications have sprung up to meet the interest in maps among the general public. The International Map Collectors' Society (IMCoS), whose first president was R.W. Shirley of the United Kingdom, holds each year a symposium in London and a meeting abroad and also runs a Journal. Map societies set up in American cities (such as Chicago and Washington, D.C.) bring together professional car- tographers, map librarians, collectors, dealers and map lovers at large. In the Netherlands, the journal *Caert-Thresoor. Tijdschrift voor de Geschiedenis van de Kartografie in Nederland* started publication in 1982. The latest journal to be launched in this area is *Speculum Orbis Zeitschrift für Alte Kartographie und Vedutenkunde*, edited by Peter H. Meurer and Dietrich Pfaehler and published in West Berlin from 1986.

In the United Kingdom, the Charles Close Society was founded in 1981 and named after an early Director-General of the Ordnance Survey of Great Britain; it can claim to be the first society internationally to confine its attention to the products of one cartographic organisation – as the editor, Yolande Hodson, remarked in the first number of its publication, *Sheetlines*. In terms of general publications designed for collectors, the Map Collectors' Series, which ran from 1963 to 1975, was edited by R.V. Tooley on the basis of fifty years' experience of the antiquarian map trade (Wallis and Tyacke, 1973). Its successor, *The Map Collector*, published at Tring from 1978, was also under Tooley's editorship and now has Valerie as his successor. This has established itself as one of the most popular and useful journals for map historians. It is also worth noting that the *Bookman's Weekly for the Specialist Book World*, published at Clifton, New Jersey, now produces from time to time Special Issues devoted to cartography, voyages, travels and exploration. This is a recognition that early maps are both of interest to bibliophiles and make good business.

These organisations and publications would have been limited in scope but for the advances in foundation activities. The development of a more professional map librarianship (Wallis and Zögner, 1979), the cataloguing of collections and the publication of catalogues have brought a wealth of source materials into the public domain. Major inventories with expert descriptive text have included Youssouf Kamal's *Monumenta Cartographica Africae et Aegypti*, edited by F.C. Weider for the prince, Lieden, 1926–53; Roberto Almagià's *Monumenta Cartographica Vaticana*, Biblioteca Apostolicae Vaticanae, 1944–55, and *Portugaliae Monumenta Cartographica*, edited by Armando Cortesao and Avelino Teixeira da Mota, Lisbon, 1960.

Such epic tasks of compilation and publication were inspired by national pride as well as scholarly devotion. The same may be said of the new series *Monumenta Cartographica Neerlandica*, edited by Günther Schilder and published by Canaletto at Alphen aan den Rijn, of which the first volume appeared in 1986. Other series to be particularly commended are the facsimiles of early atlases published by Theatrum Orbis Terrarum of Amsterdam with learned introductions, many by R.A. Skelton, and the facsimiles of both maps and atlases published by Harry Margary at Lympne Castle, Kent, complete with their scholarly apparatus.

It may be said in this context that cartobibliography, a term coined by Sir Herbert George Fordham (1928), has become a powerful tool for the use of historians of cartography. Notable examples of cartobibliographies are R.A. Skelton's *County Atlases of the British Isles 1579–1703* (London, 1970), Donald Hodson's sequel *County Atlases Published after 1703*, vol. 1 (Tewin, 1984), and Cornelis Keoman's *Atlantes Neerlandici* (Amsterdam, 1967–1985). Coolie Verner in the United States and Canada, Robert W. Karrow at Chicago, and Vladimiro Valerio of Italy, amongst others, have attempted to refine the guidelines for procedures in classifying, listing and describing early maps. A working group in cartobibliography was set up for this purpose at the Eleventh International Conference on the History of Cartography held in Ottawa in July 1985.

Exhibitions have also contributed much to the understanding of early maps and have brought newly discovered items into public knowledge. None has been more ambitious than that held at the Centre Georges Pompidou in Paris in 1980, entitled 'Cartes et figures de la terre'. This was organised by Giulio Macchi and was accompanied by a fine illustrated volume. Other encouraging developments have been the founding of lecture series, institutes and professorships. The Kenneth Nebenzahl Jr lectures on the history of cartography established on 3 June 1964 as a regular series at the Newberry Library, Chicago, have treated many themes. They have led to important publications such as R.A. Skelton's *Maps. A Historical Survey of their Study and Collection*, Chicago and London, University of Chicago Press, 1972, 1975, and *Five Centuries of Map Printing*, edited by David Woodward, 1975, and *Art and Cartography*, edited by David Woodward, 1987.

The Hermon Dunlap Smith Center for the History of Cartography was founded at the Newberry Library as a research institute in 1970. In 1958, the Agrupamento de Estudos de Cartografia Antiga created a Section on Studies of Early Cartography at Lisbon, under the direction of A. Teixeira da Mota and, in 1960, another such Section under the direction of Armando de Cortesao at the University of Coimbra. Luis de Albuquerque more recently became director of what is now called the Centro de Estudos de Cartografia Antiga. In the Netherlands at the University of Utrecht, Günther Schilder has the distinction of having held since 1981 a personal professorship in the history of cartography by appointment to the Queen.

A number of publishing achievements and projects must also be added to this list. The *Lexikon zur Geschichte der Kartographie den Anfängen bis zum Ersten Weltkrieg (Dictionary on the History of Cartography from the Beginnings up to the First World War)*, edited by Ingrid Kretschmer, Joannes Dörflinger and Franz Wawrik and published at Vienna in 1987, provides an admirable encyclopaedia. In the United States, the first of six volumes of the University of Chicago Press's *History of Cartography*, edited by J.B. Harley and David Woodward, appeared in 1987. It covers cartography in prehistoric, ancient and medieval Europe and the Mediterranean and provides a magnificently researched, documented and illustrated account by a team of experts. The further volumes are eagerly awaited. Another European work to note in the field of general cartography is that by Aleksey V. Postnikov, published in Moscow in 1985. Both Postnikov and L.A. Goldenberg in the USSR have played leading roles in work on historical cartography, both domestic and foreign. Recent classics which should also be included in this record are A.H. Robinson's *Early Thematic Mapping in the History of Cartography*, Chicago, 1982; R.W. Shirley's *The Mapping of the World. Early Printed World Maps 1472–1700*, London, 1983; Michel Mollat du Jourdain and Monique de la Roncière's *Les Portulans. Cartes marines du XIIIe and XVIIIe siècle*, Fribourg, 1984 (English language edition: *Sea Charts of the Early Explorers 13th to 17th Century*, New York, 1984); and Walter W. Ristow's *American Maps and Map Makers: Commercial Cartography in the Nineteenth Century*, Detroit, 1985.

To turn to a very different cartographic tradition, Joseph Needham of Cambridge and his Chinese colleagues and also Akio Funakoshi of Nara Women's University in Japan have documented the long and rich history of Chinese mapping. The discovery of the Han maps (c. 168 BC) in the Ming tombs was one of the great finds of this century and studies of these are in progress in China. In defining the remarkable achievements of Jesuit missionaries in China, Father Pasquale M. d'Elia S.J. has led the way. Distiguished historians of Japanese and world cartography in Japan include Hiroshi Nakamura, Nanba Matsutaro, Muroga Nobua, Unno Kazutaka and Shisue Aihara, as well as Sir Hugh Cortazzi, recently British Ambassador in Japan. Little has been published on indigenous cartography of the Indian subcontinent, so Susan Gole's current work is to be welcomed.

The range of research in the history of cartography is revealed in the *International Directory of Current Research in the History of Cartography and in Carto-Bibliography*, first published in 1975, 5th edition, Norwich, 1985.

RECENT DEVELOPMENTS

Reviewing the progress of recent years, we can say that what has been called the 'ethno-centred' character of earlier literature, with its focus on European mapping, has been rectified. The work of Malcolm Lewis of Sheffield University (UK) and Conrad Heidenreich of York University (Canada) on North American Indian mapping and that of Catherine Delano Smith on prehistoric maps illustrates this widening view. P.D.A.

Harvey's *The History of Topographical Maps. Symbols, Pictures and Surveys*, London, 1980, is one of the most useful general histories to encompass the full cultural range and reveal deeper psychological perspectives. In it, the map is seen as a means of communication, with its own language and the social context of mapping and map publishing is treated as a relevant theme. Study of the iconography of early maps has detected an implicit message to be understood in terms of the artistic, spiritual and political conceptions of the time.

Despite these widening horizons, the discipline of historical cartography has become a matter of some debate in recent years. J.B. Harley, M.J. Blakemore and Denis Wood have cast critical eyes on the philosophical principles behind the work of past and present historians: a more rigorous methodological approach is demanded. They regard the influence of librarians and collectors as having a limiting effect.

Whatever the philosophical principles in our historical studies, early maps certainly command, on occasion, the public stage. The Vinland Map, claimed to date from about 1440 and to show America before Columbus, became front page news in 1965 and caused riots among the Italians of New York. The debate on its authenticity is still going on. A Mercator atlas with autographed manuscript maps made the record price of £340,000 in Sotheby's sale rooms in London on 13 March 1979 and has since disappeared from view. Thus politics and commerce enter our field as much as any other, while scholars pursue their more sober tasks of unravelling the past. Let us end with A.H. Robinson's maxim, as we consider the infinite variety and fascination of map-making through the ages: 'There are few results of man's activities that so closely parallel man's interests and intellectual capabilities as the map'.

REFERENCES

Blakemore, M.J. and Harley, J.B. (1980) Concepts in the History of Cartography. A Review and Perspective. *Cartographica*, 17.

Gough, Richard (1780) *British Topography*, London, 2 Vols.

Fordham, Sir Herbert George (1928) *Hand-list of Catalogues and Works of Reference relating to Carto-Bibliography and Kindred Subjects for Great Britain and Ireland*, Cambridge University Press.

Harley, J.B. (1986) Imago Mundi: the first fifty years and the next ten, *Cartographica*, 23, 3, 1–15.

Harley, J.B. and Woodward, David (1987) *The History of Cartography. Volume One. Cartography in Prehistoric, Ancient and Medieval Europe and the Mediterranean*, Chicago & London, University of Chicago Press.

Ormeling, F.J., Sr (1987) *ICA 1959–1984, The First Twenty-Five Years of the International Cartographic Association*, Enschede, International Cartographic Association.
Robinson, A.H. and Wallis, H. (1987) (eds.) *Cartographical Innovations. An International Handbook of Mapping Terms to 1900*, Tring, Map Collector Publications (1982) Ltd., in association with the International Cartographic Association, 1987.
Skelton, R.A. (1972, 1975) *Maps. A Historical Survey of their Study and Collecting*. Chicago and London, University of Chicago Press.
Wallis, Helen and Tyacke, Sarah (1973) *My Head is a Map. Essays and Memoirs in Honour of R.V. Tooley*, London, Francis Edwards and Carta Press.
Wallis, Helen and Zögner, Lothar (1979) *The Map Librarian in the Modern World. Essays in Honour of Walter W. Ristow*, München, K.G. Saur.

EDUCATION AND TRAINING IN CARTOGRAPHY

Ferjan Ormeling

INTRODUCTION

On September 29th, 1981, cartographers in The Netherlands convened in Utrecht on the occasion of Cor Koeman's farewell address. As his retirement was a milestone in the development of Dutch post-war cartography, the occasion was seized upon to set future goals for Dutch cartographic education and to present these in the form of a seminar entitled 'The mapping of The Netherlands till the year 2000 AD'. Representatives of topographic and thematic mapping agencies presented their forecasts for this period and, in the ensuing discussion, the educational consequences of these projected developments were reviewed (Ormeling, 1981). Even if Koeman did not draw the blueprints for these developments, at least he had provided a number of the 'building blocks' in the 25 preceding years in which he had directed the cartography section of Utrecht University.

Nearly all the mapping agencies represented on that September afternoon had already begun the construction of data bases. Projects for speeding the provision of topographic maps were under way, and other ways of making available new topographic information were discussed. The Geological Survey and the Soil Survey described the requirements which they anticipated would be expressed by their future customers. These included the need to provide selections from the digitised soil and geological maps, plus interpretations or simulations based on the digitised data. Planning agencies like the Land Consolidation Survey and the State Physical Planning Agency described map use experiments, the results of which would soon be implemented in order to improve map readability.

Future cartographers, they emphasised, should learn to help customers to specify their needs accurately and give concrete form to their clients' wishes – the results should be aesthetically pleasing, reliable and customer-oriented. It was also envisaged that the activities of future cartographers would be more integrated with those of photogrammetrists, geodesists and remote sensing specialists than is common at present.

In retrospect, these contributions were conspicuous because the term 'Geographical Information System' was still lacking. The systems were there, however, albeit sometimes in an embryonic state and their use was foreseen. Apart from that apparent omission, these discussions could have been held today.

A DECADE OF EDUCATIONAL DISCUSSIONS

From 1980 onwards, there has been a marked increase in the number of institutionalised discussions, during seminars or special conference sessions, on cartographic education. The reason why this rather obscure Dutch gathering, only of local importance, is mentioned above is that it was the first manifestation of a new ICA policy. This policy was set in Tokyo and was directed at the production of new syllabi adjusted to future developments in our profession. A number of similar seminars or conferences will be described in this section. All of this review was begun by Karl-Heinz Meine in Tokyo in 1980.

Tokyo: Traditional Skills vs Computer Training
The Tokyo ICA conference was marked, as far as education and training were concerned, by the battle between those who advocated the inclusion of computer graphics and remote sensing in the cartography curriculum (at the cost of traditional subjects like projections and reproduction) and those that wanted to hang on, principally, to what they considered to be the core subject of our discipline – cartographic design.

Koeman's chairmanship of the Commission on Education ended in Tokyo. By then, he was worn out by the attempts to get an international team of authors to comply with the agreements made for delivering their texts for the Basic Manual of Cartography. In the preparation of this manual, the discussions on the requirements for future education had covered whether or not to include new techniques; in the event, a traditional choice was made by Koeman and his colleagues and map use and remote sensing were excluded.

During the conference, Karl-Heinz Meine took over the Commission in Education, rechristened the Commission on Continuing Education and Training. Unencumbered by the duty to finish the manual, he was free to organise a number of seminars on the new curricula, especially in developing countries. These were held in 1984 in Rabat, Morocco and in 1986 in Wuhan, China. Members of his Commission also took part in three other educational seminars, namely in Yogyakarta, Indonesia and in Paris, France, as well as the one in Utrecht. In addition, other independent seminars were organised, e.g. in Enschede and London. The results of these seminars help us to define the future needs for cartographic education and therefore also the future tasks of the Commission on Education and Training (CET), as Meine's group was called from 1984 onwards.

ITC: Needs of Developing Countries
On the occasion of the retirement of Ferdinand J. Ormeling (Sr) from ITC on April 1st, 1982, a seminar was held in Enschede on the needs for mapping and education in cartography in developing countries (Bos and Kers 1982). A crucial aspect of cartographic education was tackled by Bos – the quantification of the need for cartographers in the years to come. Based on European standards and extrapolating from Brandenburg's data, he argued that Africa alone would need some 19,000 cartographers. As African population density was much lower, however, more conservative estimates yielded the predicted need for 2,500 cartographers. Even to maintain the strength of the existing cartographic work force, some 400 cartographers should be trained each year; this number should be extended to 600 if new tasks were envisaged.

London: Education for the Future
October 18th, 1983 was the date of a seminar organised by the Cartography Subcommittee of the British National Committee for Geography (the ICA affiliating body in the UK) on cartographic education for the future. One of the main contributors was Dr M.J. Jackson of the Thematic Information Service of the Natural Environment Research Council. He had analysed the current involvement and need for cartographers in the various stages of spatial data handling. According to his analysis, the role of cartographers in the data gathering and data pre-processing stages would decline. Highly trained cartographers would, however, be needed in the data management stage but cartographers would only play the role of software providers and consultants to their users in the data-analysis stage.

As in Enschede, the question of estimating the national needs for qualified professional cartographers was raised. A newly expressed concern
was that of providing mid-career re-training. As the industry was seeking
cartographers who were prepared for immediate production work, finding
sufficient opportunities for students to receive in-house training became
crucial for educational establishments.

Rabat: State of the Art

Meine's Commission, together with the Direction de la Conservation fon-
cière et des Travaux topographiques, held a seminar in Rabat, Morocco
in April 1984 under the title 'Education of Cartographers' (Formation
des cartographes). In one week, all aspects of contemporary cartographic
education were presented in 23 lectures (Bertrand, 1985), given by a team
of experts flown in with the support of national organisations and
UNESCO, plus local experts. A distinction was made in Rabat between
educational programmes for practical cartographers (technicians and
technologists) and for scientific cartographers (engineers and scientists).
For practical cartographers, an exhaustive survey was presented of current
techniques, from hand-lettering to computer fonts and from stripping films
to scanning. In particular, production methods and production control
were described as were generalisation and relief representation, scribing
tools, colour separation and colour proofing methods. For the training
of scientific cartographers, projections, semiology, cartographic
information, toponymy, computer-assisted methods of map-design, the
application of remote sensing and map-use methods were discussed. These
map-use methods were then demonstrated during a one-day field trip into
the Atlas mountains.

Links with Remote Sensing

Cartographers at Yogyakarta's Gadjah Mada University (Indonesia)
profited from the fact that the ICA venue of August 1984 was relatively
nearby (in Perth, Western Australia); for this reason, they organised a
post-conference seminar on the integration of remote sensing with cartographic education. The major proponent of such an integration in Meine's
Commission, Richard Dahlberg, delivered a paper to this seminar.
Another important contribution was from the International Institute of
Aerospace Surveys and Earth Sciences (ITC) which presented the in-
service training package developed for cartographic draftsmen in developing countries. As in Morocco, local experts described the educational

situation in the host country and the capacity of the training establishments – such as the Laboratory of Remote Sensing (PUSPICS) in Yogyakarta.

Map Use in Paris

In 1985, shortly before the ICA Commission's meeting in Hundested, Denmark, the French National Committee for Cartography organised a seminar on the situation of cartography in mass education and its future perspectives. This seminar was sponsored by the Institut Géographique National and UNESCO. Its starting point was the widespread cartographic ignorance in both primary and secondary education, and the limited use of maps as pedagogical tools. French cartographers looked wistfully at Britain, where Balchin and Coleman (1965) and Boardman (1983) had published extensively on the need for graphicacy – but largely in vain, according to Board (Colloque International, 1985, 1986). The French themselves are leading the research on the propagation of graphic semiology to primary school pupils. Other important issues raised here were the training of teaching staff, especially in computer-assisted cartography, and the development of cartographic software for secondary schools. Though the participants stemmed mainly from France, the recommendations from the seminar – especially those regarding the improvement of cartographic education in schools – had an international relevance.

Wuhan: Advanced Cartographic Education and Training

Joel Morrison's keynote address in Wuhan – 'How to teach cartography in the future' (Proceedings, 1986) – described the future cartographer as an information broker in high precision topography and in the spatial relations to be derived from this topographical information. Other speakers aired similar views, stressing that cartographers should be aiming at the communication of spatial information and that this also implied teaching students how to organise spatial data. Knowledge of data structures, in order to implement data models in computers, was deemed necessary – as was the preparation of students for a GIS environment with powerful micro-computers and training on large data sets so as to gain experience with data storage and data base management (input, retrieval, up-dating and archiving) techniques. Despite all this additional material, the basis of the curriculum would remain the same: concern with scale, selection, projections, design, coordinate systems, map use and map revision.

Dahlberg emphasised that a fuller development of spatial theory would help to direct the development of spatial data models to represent spatial concepts, and that more work on the characteristics of spatial data should be carried out. Lack of a coherent theory of spatial relations made it difficult to design efficient data bases, to phrase queries put to data bases in an effective way, to interconnect the various subsystems efficiently and to design effective and efficient data processing algorithms (Proceedings, 1986). This lack of theory looms particularly large as very large spatial data bases begin to be developed. The message, then, was also to train students so that they could absorb continuing developments in GIS theory.

1987
For the Thirteenth International Cartographic Conference at Morelia in Mexico, a special session was arranged on education within the various sub-fields of our discipline. Education in (advanced) cartographic technology, in tactual mapping, in the history of cartography, in GIS, population cartography, environmental mapping and atlas cartography were discussed and, in total, should have given the audience an overall idea of the educational requirements as they are currently perceived by cartographers.

But related fields have also set out programmes for future education and, since cartography, photogrammetry, geodesy and geography are all converging upon Geographical Information Systems, their plans are also relevant to us. In 1987, for instance, Canadian Remote Sensing specialists organised a seminar on 'Education for the future' (Waterloo, Ontario, June 1987).

THE ROLE OF THE COMMISSION ON EDUCATION AND TRAINING

The ICA Commission on Education and Training was instrumental or centrally involved in the organisation of most of the meetings described above. Meine proved to be an able organiser, as he also succeeded in convening a large number of members of his Commission in Frankfurt (1981) and Visegrad (1983) by procuring the help of public and private sponsors. It became apparent during his chairmanship that, in order to finance the Commission's activities, it was less and less effective to appeal for support to the organisations employing Commission members; increasingly, national and international organisations had to provide support for individual participation in the Commission to be feasible. Thus, for the seminars organised in Rabat and Wuhan, fund-raising on a national level became a prerequisite for the success of these endeavours. This was in line with the experience gained at other seminars in this period, organised under the ICA Third World Seminar Program. The risks of outside support, however, are that it can be cancelled at short notice, as was the case with the proposed seminar in Munich in 1987, on computer-assisted cartographic education and training. As a consequence, the meeting had to be abandoned.

Apart from his organising talents (which were also applied for the benefit of other ICA commissions, such as for the Hamburg conference of the Commission on Cartographic Communication, as well as for improving the status of cartographers in the FRG), Meine was an inspired bridge-builder between East and West. This can be deduced from the composition of his Commission. In its meetings at Frankfurt, Visegrad, Hundested and Prague, this Commission developed the programmes for the seminars and worked on the production of an exercise manual to accompany the Basic Manual of cartography. An inventory was also made of the availability of and needs for educational material, such as text books, demonstration sets, films, slides and software.

INDEPENDENT CONTRIBUTIONS

Not all contributions to the field of cartographic education and training were channelled through CET. A number of worthwhile suggestions made outside the Commission are described below.

Monmonier was the first to draw the attention of the educational community to the fact that the map-making effort is not controlled by cartographers but by organisations and that – instead of maps – information is the principal product of mapping organisations. The major constraints on map use are now institutional and political, instead of perceptual and cognitive. This fact requires the addition of a (mapping) policy orientation to cartography courses and a focus on institutional aspects of information management (Monmonier, 1982).

A member of Meine's Commission, Fraser Taylor, simulated a carto-graphic seminar on his own by inviting a number of leading educational specialists, especially from North America and Nigeria, to comment on education and training in contemporary cartography; their contributions were collected in the series *Progress in Contemporary Cartography* (Taylor, 1985). Together with the Rabat and Wuhan seminars, the pub-lication of this book is a major educational event of this decade and since (unlike the proceedings of the two seminars) it is in English, it therefore has had a great impact.

Taylor stated that, since there are revolutionary changes in cartography because of new developments in computer technology and in communication, education in cartography must respond to these new challenges. 'The New Cartography' he discerns deals primarily with information. In his contribution to Taylor's volume, however, Keates put the readers' feet back on firm ground again by stating that – even in tomorrow's society – information will have to be extracted from maps through visual perception. Morrison's contribution to the same book urged educators not to concentrate courses too much on small-scale thematic cartography, as the majority of jobs are to be found in large-scale topographic mapping.

The five Nigerian cartographers who contributed to the Taylor book were especially concerned with the way mapping and remote sensing technology was employed to generate natural resources information, and with the part of Africans in acquiring these data. Important constraints in Africa are the existence of different geographic referencing and classification systems, incompatible resource definitions and spatial and temporal inconsistencies in resource data. As resource maps can only be successfully employed if the techniques and procedures used are completely understood by local users, the training of such local staff and the numbers in which they should be trained were important issues in the book.

Jenks, in the special *Cartographica* issue published to mark his retirement (Gilmartin, 1987), stressed that cartography students should be trained first in defining the goals for a map and in visualising the final map and its effects, before starting the physical process of map production. As cartography is an intellectual exercise and the selection of data for the map and its generalisation should be reasoned out beforehand, the actual mapping should be postponed till the student has acquired a clear idea of what he or she is doing.

ELEMENTS OF A NEW CURRICULUM

Based on the work of CET, on the independent contributions mentioned and on the discussions held during the seminars listed above, a number of requirements for a new cartography curriculum can be listed here. The role of the computer will not be mentioned explicitly, as it will be considered a fixed component of current education.

DATA CHARACTERISTICS ANALYSIS

Fig. 7.

Spatial Concepts

Since our task is to communicate spatial information and to learn about the distributional characteristics of geographic phenomena through mapping them, a prime aspect of cartographic education must be to convey knowledge of spatial concepts. Without it, we are bound to make wrong decisions in selecting mapping methods or map analysis methods and designing cartographic expert systems would be impossible.

These spatial concepts – if possible combined together in a general spatial theory – should also be taught to geographers; thus they could be a useful common start for education in our related professions. Spatial characteristics of phenomena will enable us to derive data characteristics with more ease (see Fig. 7).

Data Acquisition

Cartographers need to know as much about remote sensing (including photogrammetry) as is necessary to enable them to locate and map the resulting data and to converse with specialists about the procedures followed and their influence on spatial characteristics. Apart from that, sufficient information on the traditional aspects of fieldwork need to be taught, in order to ensure students understand concepts like representative samples, accuracy, currency and completeness of the data with which they are dealing.

One of these traditional aspects of data acquisition is the gathering of locational information; the prowess of cartographers ranges between the

Fig. 8.

ability to map surveyors' fieldwork and the construction of small-scale maps according to whatever projection that is appropriate to the map use goals (see Fig. 8). This ability of cartographers should be extended to the handling of remotely sensed data as well.

Data Processing

Acquiring insight into data processing and analysis techniques (especially for integrating remote sensing and digital cartographic data) and into the value and nature of the resulting data is a necessary curriculum part which is linked to spatial concepts. Whenever one combines, for instance, population numbers with area values, the result is a spatial ratio which can be expressed by isolines, by the choropleth technique or by proportional symbols ('average area available for every person'). Various possible definitions of the subject matter are reflected in these mapping methods: the ratio of persons to area can be understood as the number within a certain range around any point, or the number within a specific enumeration unit, divided by its area value. Different combinations of data categories lead to different spatial characteristics which, in turn, require different modes of cartographical presentation.

Design

As more and more laymen will be able to map spatial data in a CAC environment without the intervention of cartographers, students should

CARTOGRAPHERS SHOULD PLAY A MEDIATING ROLE

COMPUTER SCIENTIST

CARTOGRAPHER

SYSTEM USER

Fig. 9.

learn to mediate between computer scientists developing packages and users (see Fig. 9). This ensures that the students must be able to speak the languages of both groups. Not only should the basics of design be taught properly to cartography students, but they should also be made aware of how to express these basic tenets in a form that can be incorporated into mapping packages (Van Elzakker and Ormeling, 1984). In order to effect this, they should learn to define critical points in the design process and restrict the number of options in order to lead users to acceptable map products (see Fig. 10).

Design education also consists of learning the effects of mapping methods, for instance the visual impact of extreme values for small populations who are living in large marginal areas and are represented on choropleth maps (Census Research Unit, 1980). Other examples include the enhancement of urban areas by proportional circles and of rural areas by area symbols and the faulty effects possible from use of 3D symbols or the non-effects of piegraphs. The design stage, then, should also contain some instruction on how to inquire into the visualisation needs of one's customer (see Fig. 11), which might even go as far as role-play and discussion techniques. Based upon these visualisation needs, cartography students will have to learn how to select map methods (see Fig. 12) in the same way as users will have to be discouraged from using methods not suited for their visualisation needs.

As a special aspect of map design, atlas cartography courses should

Fig. 10.

Fig. 11.

teach the extra complications from combining maps into atlases. Issues here include standardisation and the selection of relevant map themes (see Fig. 13) so that a narrative will emerge and be suited to the information transfer needs.

Fig. 12.

Fig. 13.

Production and Reproduction

Actual design experience can be gained both on a screen and on paper; traditional manual skills will have to be de-emphasised in cartographic teaching, as will lettering and inculcation of experience in traditional reproduction techniques. Whilst the skills learned should correspond with the occupational practice, learning these manual skills tends to become too time-consuming in an environment where so many aspects of the new technology should be mastered. Moreover, in such a computer environment, more time is needed for planning and processing than for the actual mapping operation.

Information Policy and Distribution
Society's demand for spatial information, the current supply of it and the manner in which this supply is controlled for different information categories is something that needs to be taught because it will have a direct impact on cartographic careers. It is perhaps unrealistic, because of our small numbers, to think that it will be possible to change government information policies, but at least one can teach how to make people realise the consequences of these policies for the provision of spatial information. The monopoly of the state, as the prime collector and provider of spatial information, leads to many maps not being produced or, once produced, not being distributed.

But cartographers are just as guilty for the loss of spatial information because they rarely document their products and only began to store map titles and descriptions in data bases at a relatively late date. Procedures for the documentation and the storage and retrieval of bibliographic information on maps need to be taught.

Map Use and Geographical Information Systems (GIS)
As cartography's stepchild, it is no wonder that map use has been developed by non-cartographers. But as a result, this negligence is now threatening our part in the development of Geographical Information Systems. Systematic training in map use never was a part of cartographic training and this situation should be rectified as soon as possible. Based on the knowledge of spatial concepts, the various map use and map analysis tasks should be discussed with the students. This must include the possibilities for processing data, the effect of combining data categories, the nature and value of the derived data, the propagation of errors and the accuracy and scope of the final results.

Geographical Information Systems are powerful tools for data analysis and decision support but, in the hands of the uninitiated, the value of their statements is overestimated – to put it mildly. Cartographers have now to be trained to assess and explain, to correct and direct the efforts of those not familiar with the combining of spatial data.

FUTURE TASKS OF THE ICA COMMISSION ON EDUCATION AND TRAINING

The implementation of the elements described above in cartography curricula all over the world would be the appropriate long-term term of reference for CET, and all other terms of reference should be derived from it. For the next four years, this long-term aim can best be brought nearer by organising seminars on the topics that are most needed, as teachers themselves have to be trained in them. I refer here specifically to Geographical Information Systems and to the integration of remote sensing (advanced technology) in cartography curricula. Together with map use, these are the themes that are conspicuously lacking in the Basic Manual produced by CET, *Basic Cartography for Students and Technicians* (ICA 1984, 1988). As an update for this manual is sorely needed, the material from these seminars could well be used to supplement it. Such a set-up would have the advantage that there would be no need to wait for authors to substantiate their promises: the material needed would be available from the seminars themselves.

If the use of Geographical Information Systems is regarded as a specific, advanced form of map use, then one of the seminars could tackle both of these themes. It could also cater for the needs of those who require to understand spatial relationships and how to relate their work to spatial theory. For such a seminar, some form of co-operation with ICA's Commission on Map Use would be necessary. Dahlberg, who established the cartographic subject-matter needs for seminars on behalf of CET by way of a questionnaire (Dahlberg, 1987), has also suggested map use and advanced technology as the seminar themes evoking most interest amongst cartographic teaching staff.

Meine's first employer impressed upon him that, in order to ward off tomorrow's disasters, one should map today. We may paraphrase this statement and claim that, in order to improve our world, we should train cartographers to tackle tomorrow's problems with the force of tomorrow's techniques and so help to contribute to the future management of our planet.

REFERENCES

Balchin, W.G.V. and Coleman, A.M. (1965) Graphicacy should be the fourth ace in the pack. *The Times Educational Supplement*, 5 November.

Bertrand, R.J.M.J. (Ed.) (1985) La formation des cartographes. Compte rendu du séminaire de Rabat, Maroc, 1984. ITC, Enschede, 254 pp.

Boardman, D.J. (1983) *Graphicacy and geography teaching.* London: Croom Helm.

Bos, E.S. and Kers, A.J. (Eds.) (1982) Special cartography issue, *ITC Journal,* 1982–3.

Census Research Unit (1980) *People in Britain: a census atlas.* London: HMSO.

Colloque International Education et Cartographie (1985) *Bulletin du Comité Français de Cartographie,* 106,107, 1985,1986, 123pp.

Dahlberg, R.E. (1987) An ICA response to the educational challenges of cartography in transition. *Proceedings 13th International Cartographic Conference of the International Cartographic Association,* Morelia.

Gilmartin, P.P. (1987) Studies in Cartography. A festschrift in honor of George F. Jenks. *Cartographica,* 24, 2, 135pp.

International Cartographic Association (1984, 1988) *Basic Cartography for students and technicians.* 2 Vols.

Lawrence, G.R.P. (Ed.) (1984) *Cartographic education for the future.* London: British Cartographic Society, 96pp.

Monmonier, M.S. (1982) Cartography, geographic information and public policy. *Journal of geography in higher education,* 6, 2, 99–107.

Ormeling, F.J. (Ed.) (1981) De kartering van Nederland tot het jaar 2000 (The Mapping of the Netherlands till 2000 AD). *Bulletin van de vakgroep Kartografie* No. 13, November 1981, Utrecht, 62pp.

Proceedings First WTUSM/ICA Seminar on Advanced Cartographic Education and Training (1986) Wuhan: Wuhan Technical University on Surveying and Mapping (3 volumes in Chinese).

Seminar on Professional Education and Training in Cartography and Remote Sensing (1984) Yogyakarta: PUSPICS, Gadjah Mada University, Indonesia, 138pp.

Taylor, D.R.F. (1985) *Education and Training in Contemporary Cartography.* Volume 3 of the Progress in Contemporary Cartography Series. New York: Wiley, 324pp.

Van Elzakker, C.P.J.M. and Ormeling, F.J. (1984) Computer-assisted statistical mapping systems: user requirements. *Technical papers of the 12th Conference of the International Cartographic Association,* Perth, Australia, 1984, pp.535–553.

CARTOGRAPHY AND THE DEVELOPING NATIONS: SOME NEW CHALLENGES

D. R. Fraser Taylor

INTRODUCTION

Ferdinand Ormeling, to whom this volume is dedicated, has worked extensively in developing nations as his biography shows (see the introductory chapter by Hedbom and Böhme). Through his work at ITC, he has contributed to the education and training of cartographers from all over the developing world. While President of the ICA, Ormeling also made major efforts to involve the developing world in the work of the organisation, with considerable success. It is therefore appropriate that a chapter on cartography in the developing world be included in this volume. This chapter will begin by looking very briefly at the history of cartography in the Third World before considering the new challenges facing a cartography for development.

THE HISTORY OF CARTOGRAPHY IN DEVELOPING NATIONS

The term 'Developing Nations' used in this chapter is a relatively recent one and some would argue that it is not a particularly accurate one. For several centuries many of the countries now placed in this category were colonies of European powers such as England, France, Spain, Portugal, Belgium, Germany and (Ormeling's native) Holland. Others, although not directly colonised, were strongly influenced by the powerful imperial systems which grew up – especially between 1500 and 1900 AD. Some historians have argued that, during this colonial period, the socio-economic system was as much one of exploitation and domination as it was of 'development'. Development suggests that, from the perspective of the inhabitants of the colonial empires, positive change was taking place. Several authors such as Samir Amin, Walter Rodney, and André Gunder Frank have argued that the process was as much one of 'under-developing' as 'developing'. In Africa, for example, it is difficult to see the removal of millions of people as slaves as a positive development for that continent – although their labour helped to build the economies of both the USA. and England. Regardless of what judgement is made on such issues, there is no doubt that many of the maps of Asia, Africa and Latin America made by European powers were initially for the purpose of exploration followed by control and exploitation. This is certainly true of the mapping of Africa, for example – especially in the 19th century when the 'scramble for Africa' among the colonial powers of Europe took place. The map was a major tool at the Conference of Berlin, which began the process of the definition of 'spheres of influence'. The major colonial powers proceeded to draw lines on the map and these resulted in the creation of boundaries which took little cognisance of the realities of the African societies of the time. It was not unusual for such boundaries to run right through the centre of the territory of an ethnic group and, in some instances, members of the same group found themselves arbitrarily divided among several new colonial nations.

From an African perspective, the lines drawn on the map as a result of the processes set in motion by the General Act of the Conference of Berlin of 1885 were illogical, but from the perspective of the quarrelling, competitive colonial powers they represented a complex compromise to a conflicting set of geo-political, strategic, religious and economic goals. The Act began the 'great game of scramble'. 'For the Europeans it *did* become a gigantic game, some super "Monopoly", played with real land and people. Zanzibar was traded for Heligoland, Cameroun became Kameroun for "a free hand in Morocco"' (Griffiths 1984, p.45). In the modern era, Africa has continued to suffer from the problems created by these colonial boundaries, which have proved impossible to redraw, however logical such a redistribution of territory might be in cultural terms. The Organisation of African Unity has formally endorsed the boundaries established as a result of the Conference of Berlin and the actions which resulted from it, with all their idiosyncrasies. One of the better known of these is Queen Victoria's decision that the boundary between British and German East Africa be drawn in such a way that

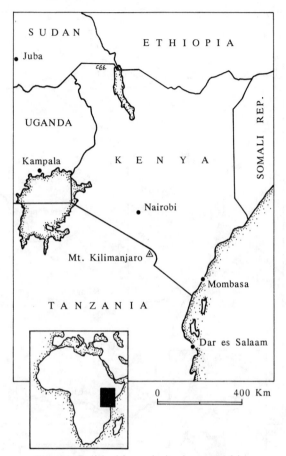

Fig. 14. Political boundaries in east Africa.

Mount Kilimanjaro, the highest peak in Africa, be part of German territory (Fig. 14). She is reputed to have made this decision as a birthday gift to her grandson, the Kaiser Wilhelm III. Another is the existence of the Caprivi Strip which resulted from German insistence on having access to the Zambezi from her colonial possession of what is today Namibia (Fig. 15).

In few areas of what is now called the developing world did a substantive indigenous cartography evolve. A major exception to this was

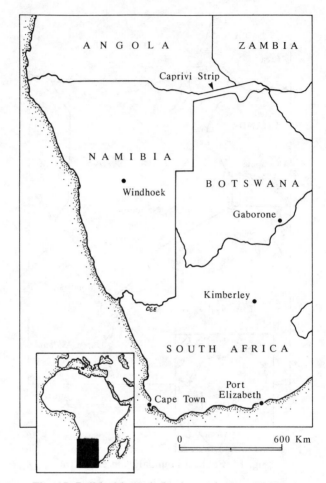

Fig. 15. Political boundaries in south west Africa.

China, where there was a cartographic tradition stretching over two thousand years. References to maps in China go back to the third century BC but the earliest maps of which we still have copies date from around 170 BC. Figure 16 shows one of the two maps found in one of the Han tombs of the period. These were hand-drawn in colour on silk. From the early beginnings, Chinese cartography developed in a quite remarkable form, as chronicled by Needham (1959, 1981). Figure 17, which is the

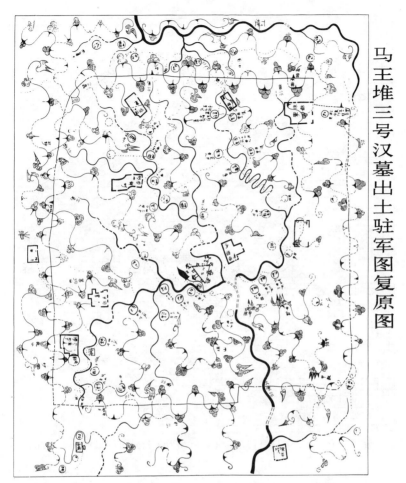

马王堆三号汉墓出土驻军图复原图

Fig. 16. The garrison map from Han Tomb Number Three.

very well known Yu Ji Tu of 1137 BC, shows an indigenous cartography far in advance of any other nation of the time, and the development of Chinese cartography continued unbroken well into the 18th century. When the Jesuit cartographers reached China in the latter part of the 16th century, the cartography they found was superior to their own. The marriage of some of the ideas of Chinese cartography with those of contemporary Europe led to some interesting cartographic developments

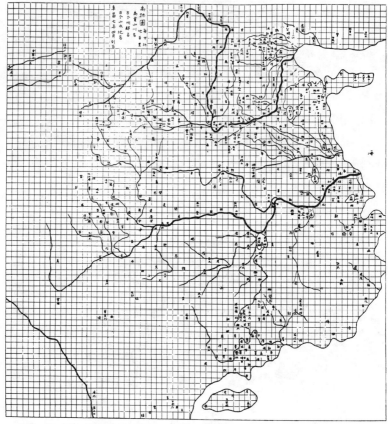

Fig. 17. The Yu Ji Tu 1137 AD (Map of the Tracks of the Great Yu).

(Wallis 1980) and the maps of 18th century China were among the best maps produced at that time. Chinese cartography went into decline in the 19th century as Chinese cartographers were unable to escape the constraints of the rectangular grid – a feature of Chinese cartography from as early as the 3rd century AD. China also suffered badly – during the 19th century in particular – from the imperialism of powers such as England, France, Germany and the United States. The Chinese experience is, however, unique: the history of cartography of most parts of the developing world is the history of the foreign powers who mapped them. In some instances, this led to the groundwork being laid for an excellent topographic base map. One such example is that of the Survey of India which was founded in 1767 and produced a truly remarkable topographic map base; this was, however, more the exception than the rule. Cartography was, in the main, the hand-maiden of European domination, exploitation and military control.

CARTOGRAPHY FOR DEVELOPMENT

The period after the Second World War saw the beginnings of a rapid political decolonisation, with nation after nation gaining independence from their former colonial masters. Independence was accompanied by a desire to increase the pace of socio-economic development and, in many nations, the challenge has become the dominant one facing post-colonial societies. Thus the demands made on a cartography for development are quite different from that of a cartography for domination and exploitation. Cartography has always had a role to play in the process of socio-economic development over the centuries but that role has rarely been the dominant one. Some of the most rapid periods of cartographic development have taken place as a result of the demands of war, and cartography has not responded nearly as well to other challenges. The challenges of a cartography for development demand new and innovative responses. Cartographers have been slow to respond to these and in many nations a military survey mentality still dominates the discipline. This can be a limiting factor in the emergence of a new and effective cartography for development.

There is no doubt that an accurate up-to-date topographic base is necessary for national development but, in itself, it is by no means sufficient. If cartography's main task is seen solely as the provision of such a base, to the neglect of other tasks, then much will be lost. The cartography of development requires major emphasis to be placed on applied thematic mapping. The thematic map must become a more important product if cartography is to make a more effective contribution to development challenges. The topographic map is a product usually in greatest demand during wars; the thematic map should be seen as a major cartographic product for another kind of war – the war against poverty.

One starting point in encouraging the changes required is a redefinition of cartography. The most widely accepted definition at present is given in the ICA Multilingual Dictionary and reads as follows:

'The art, science and technology of making maps, together with their study as scientific documents and works of art. In this context maps may be regarded as including all types of maps, plans, charts and sections, three dimensional models and globes representing the Earth or any celestial body at any scale.' (ICA, 1973)

Technological development has led to considerable change in cartography and a more recent description of the discipline reads:

'. . . an information transfer process that is centred about a spatial data base which can be considered, in itself, a multi-faceted model of geographic reality. Such a spatial data base then serves as the central core of an entire sequence of cartographic processes receiving various data inputs and dispersing various types of information products.' (Guptill and Starr, 1984, p. 1)

Neither of these definitions captures the central thrust of cartography for development. One formal definition which comes close is that given by the Wuhan Technical University of Surveying and Mapping:

'The art, science and technology of making all types of maps by employing results obtained from surveying, reconnaisance, remote sensing and other data available for use in economic construction, national defence, international relations, education, culture, tourism, etc.' (Wuhan Technical University of Surveying and Mapping, 1985, p. 15)

This definition places primary emphasis on the uses to which cartography is to be put, rather than on the technologies used to make the map or on the maps as a means of communication. The last two of these concepts have had an important influence on the way in which cartographers have regarded their discipline over the last decade, as several chapters in this volume illustrate.

TECHNOLOGICAL CHANGE AND THE CARTOGRAPHY OF DEVELOPMENT

There are numerous references available which describe the incredible pace of technological change taking place in cartography (e.g. Auto Carto Six 1984, Auto Carto Seven 1985, Auto Carto London 1986 and Auto Carto Eight 1987). Joel Morrison (1986) has suggested that the change is now so far advanced that little is to be gained from attaching the adjectives 'automated' or 'computer-assisted' to cartography. The influence of the computer has become all-pervasive and it is now central to our discipline. Some would argue that the core of the discipline remains essentially the same and that only the technology has changed (Kadmon 1984). It has been argued elsewhere (Taylor 1985) that this is not the case and that the changes are so pervasive that a 'New Cartography' is emerging.

There has probably never been a period in the history of cartography where the pace and impact of technological change has been so rapid and great. The result is, perhaps understandably, a mad scramble to keep up with technological change, reflected in attendances reaching the one or two thousand mark at cartography conferences such as Auto Carto London. The leading edge of the change is coming from the development of new hardware, with software lagging far behind and applications and the needs of the user – despite rhetoric to the contrary – a very distinct third.

There has been an overwhelming emphasis, especially in the industrialised countries, on an approach which can be described as the 'development of cartography'. Cartographers in academia, government and industry are absorbed with the challenges posed by technologically-forced change; developments in data capture, data base structure, data base management, software for data manipulation, etc. have been required to adapt cartography to the new technology. These are undeniably important tasks, but the sense one gets from reading the literature is that many cartographers and cartographic agencies are more concerned with the development of cartography from the perspective of the producer than from that of the user.

The 'user' has always been important to cartographers and genuine attempts have been made to determine what the user wants, so that our applied discipline can deliver useful products. Having determined what the user was thought to want, cartographers then proceeded to produce it. There was never any doubt of where control of production lay. This was clearly in the hands of the cartographers, especially those in the large national mapping agencies. In many instances the main purpose of mapping was (and still is) military security and defence – which helps to explain the need for control. Maps remain classified products in many countries even today and, if military uses are deemed to be the major ones, then such control of production and distribution may continue. Military uses will continue to influence cartography but in most countries they are by no means the only uses and, for all purposes other than that of security and defence, a change to information sharing, effective information distribution and the formation of new relationships between cartographic producers and users are required. These are, of course, also issues for military cartographers but the users with whom information is shared and to whom information has to be distributed are much more clearly defined.

A fuller acceptance of the idea that the needs of production agencies should be tempered by the needs of the user/producer should lead to changes even during the current 'development of cartography' phase in North America and Europe. For example, the data base structure and design most suited to the needs of a production agency may be quite different from that of many of the user/producers.

Recently, Boyle (1987) has been involved in research in this area where 'Attention was focussed more on the needs of the user of cartographic data than on the problems of the producer. . . .' The results of his survey, which covered users in a number of nations, are interesting and worth quoting at some length.

The investigator and many traditional cartographic producers have for long assumed that the needs are best met by the generation of digital cartographic data in vector form. In almost all the uses examined by Boyle, however, the line data from topographical maps or cadestral ones was only needed for background imagery on displays and photos and this could well be met by cartographic data in digital raster format. Parallel discussions with producers showed that raster format products could well be made in one hour as against one hundred hours for clean vector data; as equipment usage was similar, the cost differences are evident. The supply of cartographic data in digital raster format is not a new idea, but one which has to be continuously re-examined with new developments in use and particularly in relation to the availability of powerful microprocessor systems at the user's establishment. Boyle (1987) considered that cartographic data should perhaps be primarily supplied in digital raster form, with the user converting all or parts of the data to vector form as required and if necessary.

In a similar vein but from a very different perspective, Burrough (1986, p.146) has commented, 'It is only recently with the arrival of small, cheap but powerful processors and good raster display systems, that more people can work with natural resource data in ways that have not been dictated by CAD/CAM systems design'. The lack of attention paid to research on maps produced from raster data became obvious in the development of the impressive Domesday Project in Britain. Openshaw *et al.* have commented, 'The principal problem with a raster graphics choropleth mapping and GIS is that the required technology has been neglected by geographers partly due to the tremendous success of vector graphics systems. . . .' (Openshaw *et al.* 1986, p.297).

In the nations of Africa, Asia and Latin America, remotely sensed imagery – which is dominantly in raster format – plays a very significant role in socio-economic development planning. This is a role which is likely to increase, especially in countries like India and China. As high resolution satellite images such as SPOT become increasingly available, it may well be that these images may, in fact, be used as the fundamental background for thematic mapping – especially at small and medium scales. '. . . if high resolution satellite images can be the source of numerous topographic data, they can also constitute, like aerial photographs as well, the basis of the cartographic representation itself. . . .' (Denegre *et al.* 1986, p.528).

The technological developments in cartography create something of a dilemma for developing nations. Some have argued that the sophisticated techniques of computer-assisted cartography are not appropriate to the Third World context and have little relevance to developing nations. They argue that what is required is a technology which is labour- rather than capital-intensive, especially in situations where foreign exchange is scarce or unemployment is at high levels. When computer-assisted cartography was available only on large, expensive mainframes, the strength of this argument was considerable but the recent developments in micro-computer technology have created a new situation.

Appropriate technology is perhaps best defined as that most appropriate to the task at hand. There are sometimes advantages to the adoption of the most modern concepts and techniques. The development of cheap micro-computer technologies is substantially increasing the feasibility of the introduction of computer-assisted cartography in any developing countries.

These technologies are not particularly capital-intensive and may allow the cartographers of the developing world to 'leapfrog' over existing approaches in a way which will make their products of more direct utility to the development process. The newest technologies increase the productivity of labour rather than replace it, and are also particularly well-suited to the decentralised style of planning which is growing in popularity. What is required is an effective merging of new and existing techniques in a way which is appropriate both to the tasks at hand and to the socio-economic context in which the technologies are to be applied. Some encouraging steps are already being taken and a few examples will be given to illustrate this.

TOWARD A NEW CARTOGRAPHY FOR DEVELOPMENT

Judging from personal experience and from the available literature, the country which is making the most impressive steps in the application of cartography to development problems is The People's Republic of China. This provides several examples of the types of changes required to initiate a new cartography for development.

The main purpose of cartography in China is to contribute to the socio-economic development of the nation. This has been clearly and explicitly stated, as can be seen in the definition of cartography by the Wuhan Technical University of Surveying and Mapping given earlier. The responsibility for cartographic leadership in China is shared by two ministries – the Ministry of Defence and the Ministry of Urban and Rural Construction and Environmental Protection, under whose direction the National Bureau for Surveys and Mapping was placed in 1982. Cartography is explicitly mentioned in the latest five year plan (People's Republic of China 1986) and special emphasis is being given to thematic mapping by all agencies involved in map production in China at all levels. This clear sense of direction is a key element in the evolution of a cartography for development, and few nations have such explicit national policies on thematic mapping.

China is making extensive use of aerial photography and remote sensing for thematic map production. The use of remote sensing technology is an interesting example of a combination of capital and labour-intensive approaches. China has launched her own satellites and can also receive imagery from LANDSAT, SPOT and other satellites. This involves the use of high technology which is quite capital-intensive. In the interpretation of this imagery, however, a labour-intensive approach has been used. Although image analysis systems exist, most imagery is interpreted visually with a very extensive programme of field checking and verification. In part, this is due to the complexity of the imagery received, which cannot be easily analysed by machine, but it is also a deliberate policy choice by the Chinese to utilise 'brainware' rather than software or hardware.

A large-scale base map is necessary for planning in rural China – especially since the implementation of the responsibility and contract systems which involve the leasing of small areas of land to peasant farmers. The basic map scales being used for this purpose are 1:5,000 and 1:10,000 and China has mapped over three million square kilometres at this scale. This large-scale base mapping will be complete for the developed areas of China by 1990 but, in the interim, extensive use of large-scale air photographs and remote sensing imagery is being made both as a source and as a base for thematic mapping. This is also true for medium- and small-scale mapping of the mountainous and developed parts of Western China.

Computer-assisted cartography and geographic information processing are being introduced at various scales in China and, although the work is at a relatively early stage, some interesting and innovative approaches have been used which hold considerable promise for the future. Extensive and increasing use is being made of micro-computer technology. Zhang has observed, 'With the extensive application of micro-computers in the 1980s in China, the CAC based on micro-computers has rapidly grown for its major advantages of greater efficiency and lower operating costs. At present, a great number of low-grade plotters of mini or middle size are being used in China for a wide variety of purposes including geology, petroleum exploitation, geography, environmental protection, statistics and teaching. In addition to the micro-computer and the plotter, the digitiser and the graphic display device have been gradually introduced to form a full-set hardware configuration as a stand-alone micro-computer mapping system for a wide range of system engineering objectives.' (Zhang 1987, p. 48).

China's 2,300 counties are considered to be the most important administrative units in terms of production, especially, but by no means exclusively, in the field of agriculture. The importance of information for local level planning has been realised, and the plan is eventually to have a micro-computer-based information system with a mapping capability in every county. At present, only a few counties have such systems on an experimental basis. He and Yang argue that, 'The county is an important basic unit in China's administration, and hence its information system is an important component of a national REIS [Resource and Environmental Information System]. Among the systems already established, some are concerned with the management of agricultural information, agricultural planning and production and some combine agriculture, industry and commerce. Usually, they have the common functions of inquiries and retrieval, statistical analysis, resources inventory, automated mapping, simple planning, assessment and prediction.' (He and Yang 1986, p. 25–26).

The system built for Daxing County (Li and Sun 1986) has over a hundred data elements dealing with agriculture, forestry, animal husbandry, fisheries and natural resources. It has been built on an IBM-PC/

XT and the county also has a digitiser and a plotter. The system can perform functions such as search, statistical calculations, synthetic appraisal of agricultural resources and dynamic modelling and can produce the results in graphic form. The system being built for Fushui County deals with land resource information and is still at the experimental stage. The purpose of the system '. . . is to provide some information on management, exploitation and decision-making for the county government and related users' (Zhong and Zhong 1987, p.52) and the system uses optimisation analysis to select appropriate planning strategies.

Micro-computer mapping systems are also being developed for urban mapping and in April 1987 an atlas for the City of Luoyang was completed through use of a micro-computer. This was one of the products of the Local Resources Information System for the Luoyang economic zone (Sun 1987). There are also several impressive urban mapping systems being developed in China, including the system for Tianjin. The software for this system is written in Fortran IV and assembler and it was developed on a mini-computer rather than a micro. The system was developed to map the urban environment and can produce maps by line printer, plotter or on a terminal. 'The system is able to deal with environmental monitoring data, investigation and single statistic maps, human aerial photo interpretation, environmental system analysis and evaluation data. It is able to position and quantify environmental factors and brings data into the urban environmental data system to establish data files, to organise and manage data, to classify and analyse data, and to draw the six structural types of maps by computer.' (Tsui, Lin and Wang 1987, p.34).

Special purpose micro-based mapping systems are also being developed. One of the most interesting of these is that to study soil erosion in the loess soils of China (Chen 1986; Chen, Chi, Fu *et al.* 1987). There are over 600,000 square kilometres of loess soils in China and over two-thirds of this area is at risk from soil erosion. There are an estimated 300,000 small valleys larger than one kilometre in size cutting into the loess plateau and washing more and more soil into the Yellow River. One such small valley at Wanjiagou in Lishi County has been chosen for the study. The area is only 91 square kilometres in size, but data have been accumulated for a thirty-year period. A combination of remote sensing, modelling and computer mapping is being used to help understand the soil erosion processes and patterns. A detailed Digital Terrain Model has been built from the topographic base map at 1:5000 scale.

Micro-based systems are also being built at the regional scale. The system for Hainan Island covers the 18 counties found on the island and has over 30 data elements. The system has both a 100 square kilometre grid and a vector data base using county boundaries. The system was developed on a 64K Apple II and uses dBASE II (Wang, Dong, He 1985). A more recent regional system is that being built for the Luoyang Economic Zone mentioned earlier (Sun 1987). This system combines a data base on resources, a decision-making system and a graphic display system for multi-element analysis and presentation. At the provincial level in China, a *Population Atlas of Jiangsu Province* has been produced using an IBM PC and a small plotter (Sun, Wang, Tang 1987). The atlas is one product of a multi-variable geographic information system being built for Jiangsu.

At the national level, a Chinese Tourism Resource Information System is also being built on IBM PC/AT micros (Yan, Zhou and Shi 1987). The system is designed to contribute to China's growing tourist industry and maps are one of the four main data bases in the system. There are 120 data items in the system, covering a wide variety of topics of interest to tourism and, 'Besides general programs for storage, updating, retrieval and statistics, various application models are developed in the system, like the software for regional tourism evaluation, and optimum route selection for travelling, etc.' (Yan, Zhou and Shi 1987, p.44).

The introduction of a new cartography for development in China is at an early stage and is not without its problems. One of these is how best to integrate the new technologies with the old. Another is how the system of education is to be changed to meet the new challenges. Systems are also developing in China in an uncoordinated fashion and the difficult problems of the definition of standards and the connectivity of systems have yet to be resolved. An indication that the building of such systems can be quite labour-intensive, even though high technology is being used, is given by the experience of the National Agricultural Information System. Since 1979, over 400,000 people have been involved in the data collection for this system (Zhang and Kou 1987). Despite the problems, impressive progress has been made.

CONCLUSIONS

Cartography played a major role in the exploration, military domination and exploitation of many of what are today called the developing nations. The challenges now facing these nations are enormous and cartography has a role to play in their solution but it is a very different role to that of the cartography of imperialism and colonialism. The latest developments in micro-computer and telecommunication technology may present an opportunity for cartography to increase the value of its contribution. The international cartographic community should join with the cartographers of the developing world in responding to the new challenges. The approaches being used in China are an interesting step in this respect.

REFERENCES

Auto Carto London (1986) *Proceedings of Auto Carto London*, M. Blakemore (Ed.), Vols. I and II, London.

Auto Carto Six (1984) *Proceedings of the Sixth International Symposium on Automated Cartography*, B. Wellar (Ed.), Vols. I and II, Ottawa.

Auto Carto Seven (1985) *Proceedings of the Seventh International Symposium on Computer Assisted Cartography*, ASP and ACSM, Washington.

Auto Carto Eight (1987) *Proceedings of the Eighth International Symposium on Computer Assisted Cartography* (Ed. N.R. Chrisman), ASP and ACSM, Washington.

Boyle, A.R. (1987) 'Examination of Problems of Data Exchange in Mapping and Development towards the Practical Application of Information Utilities', Progress Report in Energy, Mines and Resources Canada Research Agreements Program in Natural, Physical and Social Sciences and Engineering Progress Summary, 1987.

Burrough, P.A. (1986) 'Five Reasons Why Geographic Information Systems are Not Being Used Effectively for Land Resources Assessment', in Blakemore, M. (Ed.), *Proceedings of Auto Carto London*, Vol. 2, 139–48.

Cheng Hungjin (1986) 'Building Micro-Computer GIS at Small Valley of Loess Plateau', Resource and Environment Information System No. 1, p.120, Lab of REIS, Institute of Geography, Chinese Academy of Sciences, Beijing.

Chen Hongjin, Chi Tianhe, Fu Leyuan, Zao Hui, He Jianbang, Cai Jianggao, Chen Ho (1987) 'A Study Based on Micro-Computer GIS on the Change Patterns of Soil Erosion Environment in a Small Valley on the Loess Plateau', *Proceedings of International Workshop on Geographic Information Systems*, Beijing '87 (Abstract), p.55, LREIS, Beijing.

Denegre, J., Deschamps, J.C. and Galtier, B. (1986) 'LANDSAT and SPOT High Resolution Images: A New Component for Geographic Data Bases', in Blakemore, M. (Ed.) *Proceedings of Auto Carto London*, Volume I, 527–37, London.

Guptill, S.C. and Starr, L.E. (1984) 'The Future of Cartography in the Information Age', in ICA Commission C, Computer-Assisted Cartography Research and Development Report, compiled by L.E. Starr, 1–15, Washington, D.C.

Griffiths, Ieuan, L.L. (1984) *An Atlas of African Affairs*, Methuen, London.

He Jianbang, Yang Kai (1986) 'Researches and Progress in the Resources and Environment Information System in Some Chinese Institutions: A General Survey', Resource and Environment Information System No. 1, 20–29, Lab of REIS, Beijing.

International Cartographic Association (1973) *Multilingual Dictionary of Technical Terms in Cartography*, Wiesbaden, Germany, Franz Steiner Verlag.

Li Xiang Zhi, Sun Yan (1986) 'The Research of Agricultural Information System at a County Level', Resource and Environment Information System No.1, Lab of REIS, Beijing.

Needham, J. (1959) *Science and Civilisation in China* Vol. 3, Cambridge University Press, Cambridge.

Needham, J. (1981) *The Shorter Science and Civilisation in China* (An abridgement of J. Needham's original text by Colin A. Ronan), Vol 2, Cambridge University Press.

Openshaw, S., Wymer, C. and Charlton, M. (1986) 'A Geographic Information and Mapping System for the BBC Optical Discs', *Transactions of the Institute of British Geographers* N.S.11, 296–304.

People's Republic of China (1986) The Seventh National Economic and Social Development Five Year Plan, Beijing (In Chinese).

Sun Jiulin (1987) 'The Establishment and Applications of the Regional Land Resources Information System on Micro-Computers', *Proceedings of International Workshop on Geographic Information Systems*, Beijing 87 (Abstract), p.46, LREIS, Beijing.

Sun Yamei, Wang Ruilin, Tang Qin (1987) 'Automated Cartographic System for Population Maps', *Proceedings of International Workshop on Geographic Information Systems*, Beijing '87 (Abstract), p.47, LREIS, Beijing.

Taylor, D.R.F. (1985) 'The Educational Challenges of a New Cartography' in Taylor, D.R.F. (Ed.), *Education and Training in Contemporary Cartography*, John Wiley and Sons, Chichester.

Tsui Weihung, Lin Huajiang, Wang Shujie (1987) 'Methods and Software Systems for Computer-Aided Mapping', *Proceedings of International Workshop on Geographic Information Systems*, Beijing '87 (Abstract), pp.34–35, LREIS, Beijing.

Wallis, H. (1980) 'Missionary Cartography in China', Paper read to the 11th International Conference of the ICA, Tokyo.

Wang Ze-Shen, Dong Han Fei, He Jainbang (1985) 'Experiment on Establishing Hainan Island Geographic Information System by Means of a Micro-Computer', *Proceedings of Development and Applications of Remote Sensing for Planning Management and Decision Making*, 375–90, Publishing House of Surveying and Mapping Centre, Beijing.

Wuhan Technical University of Surveying and Mapping (1985) *The Wuhan Technical University of Surveying and Mapping* (WTUSM), p.37, Wuhan.

Yan Shouyong, Shou Hairong, Shi Zhongzhi (1987) 'Chinese Tourism Resource Information System', *Proceedings of International Workshop on Geographic Information Systems*, Beijing '87 (Abstract), p.44, LREIS, Beijing.

Zhang Quiaoling, Kou Youguan (1987) 'A Study on the Information System for Agricultural Resources and Economy', *Proceedings of International Workshop on Information Systems*, Beijing '87 (Abstract), p.9, LREIS, Beijing.

Zhang, Wen-Zhong (1987) 'Micro-Computer Mapping: Its Development and Application', *Proceedings of International Workshop on Geographic Information Systems*, Beijing '87 (Abstract), p.48, LREIS, Beijing.

Zhong Saiguo, Zhong Ershun (1987) 'A Preliminary Research on Land Resources Information System (LRIS) at Fushui County', *Proceedings of International Workshop on Geographic Information Systems*, Beijing '87 (Abstract), p.52, LREIS, Beijing.

THE IDEAL MAPPING PACKAGE

Ruth Blatchford and David Rhind

INTRODUCTION

Computer-based mapping facilities have now been in use for about thirty years. In that period, considerable evolution has occurred from crude systems producing maps by using standard alphanumeric output devices, to the more flexible and general-purpose software available today. Some of the important developments have occurred outside the professional cartographic community: one example of this is the incorporation of mapping facilities in widely-used statistics packages (such as SAS). At the same time, the rapid development and spread of Geographical Information Systems (or GIS) has greatly widened the range of data available in computerised form and which may readily be mapped. The range of disciplines exploiting such mapping has greatly extended. In particular, there has been a break-down of the concept of a dichotomy between topographic and thematic mapping. The effect of all this has been to stress the generality òf the processes carried out in computer-based mapping, almost irrespective of the data source.

No introduction to a discussion on computer mapping would be complete without an acknowledgement of the fundamental effects wrought by the rapid increase in hardware performance per unit cost. This has brought what were mainframe capabilities only five years ago to the individual user of a micro-computer. It has also permitted the development of new forms of interaction between man and machine, as typified by the Apple Mackintosh. Unsurprisingly, therefore, mapping packages can take on many different forms. Their success can partly be gauged by the standard of the resulting cartography. However, as we shall see, many other criteria are involved in deciding what is the 'ideal'.

This, then, would seem to be an appropriate moment to attempt a summary of the capabilities required in a general-purpose mapping system, whether this is used as a free-standing package or simply as one part of a GIS. It is our hope that this paper will provide hints on desirable capabilities of future systems and also prevent misconceived innovations. We concentrate on functions which are *required* in an ideal mapping facility rather than whether or not they are presently available in some package or simply perform very slowly. Because of the constraints of space, most discussion is restricted to the task of map design and not extended to include modes of data entry and processing. But we start by posing and answering a simple, if fundamental, question.

WHAT IS 'IDEAL'?

It may be argued that the variety of needs of map makers and the variation in characteristics of data sets to be mapped render the concept of an 'ideal' tool totally meaningless. That view, however, is rejected for the following reasons:

—Within broad groups of data types, the same mapping techniques are already widely used (Robinson *et al*, 1984).

—Existing mapping packages such as GIMMS (Waugh and Chulvick, 1982) have been used successfully for an astonishingly wide variety of tasks. Their biggest failing at present is probably their slow performance and poor user interfaces, the former arising from their generality.

—In our experience, all users (except some individuals engaged in production work to unchanging specifications within a national mapping organisation) have need for a range of functions much wider than they currently realise and (hence) would specify if asked to fill in a questionnaire on the subject.

—Perhaps the greatest institutional and personal cost in using any large computer system is the time required to learn how to 'drive' the system competently. Typically, existing mapping packages use a wide variety of methods, often differing fundamentally, to control the tasks carried out. A single command system, logically constructed and very general, would therefore be of enormous benefit to users. In the absence of any universal 'cartographic' command language, this is best achieved by the use of one system with a wide range of capabilities.

Given all this, we believe that a single cartographic system or suite of software which provides all the facilities needed by any type of user in a 'friendly' manner and which can run on any size of data set will constitute the 'ideal'. However, we readily acknowledge a number of provisos:

—What is ideal on one computer system may be less than ideal on another,

owing to variations in the 'fit' between concepts and commands of the cartographic system and those of the host computer system.
—Different users will certainly place greater weight upon the merits of different parts of the system at different times, e.g. UK users may have less need of projection change facilities than their counterparts in the USA.
—There is evidence that traditional cartographic products are culturally-specific and, to that extent, there may well be cultural variations in what is perceived to be 'ideal'. Parenthetically, we suspect that ready availability of computer mapping packages from Anglo-Saxon sources may be diminishing this cultural diversity through 'cartographic imperialism'.
—The price of software may, in practice, influence the view of what is ideal. This is likely to become increasingly true as hardware costs fall to insignificant levels compared to those of software and of the time of the user.
—What follows assumes that current cartographic techniques continue to be seen to be acceptable to the profession and to customers for mapping.

THE FUNCTIONALITY OF THE IDEAL SYSTEM

Ignoring the data entry and validation functions (see above), we take it as self-evident that the ideal facility should provide the capability for both cartographic design and production. Used in the sense employed by Keates (1973), this therefore includes the need for tools to facilitate selection and creation of symbolism, the arrangement of these symbols harmoniously on a base and the setting out of auxiliary detail (such as title and legend) as well as data analysis and conversion and the reproduction of the graphic image.

Accepting that all these tasks need to be addressed, there are three sets of capabilities which must be present. These are the general operational, the design, and the output aspects of the mapping tool. Each is now examined in more detail.

The General Operational Aspects
The operational qualities of the ideal mapping system are set out below. Certain of these requirements overlap but each is now considered in turn.

Interactive design capability
This allows map design to take place in near real-time at the terminal. The functions required for this task are the abilities to:

—Delete either all features of an indicated type or only that one identified by pointing.

—Move features within an interactively defined window, including re-scaling (this often causes problems when hardware-generated text is used since few variations in size are available).
—Change of graphic symbolism for all features of a given type or one indicated feature (such as colour, shading, line or point symbology).
—Add, delete or change the order or value of attributes of any feature selected by pointing.
—Generate all the other conventional cartographic elements (neatline, legend, scale bar) as determined by the map designer.

'User-friendly' man/machine interface

It has shown that different users of GIS prefer different modes of issuing commands in different circumstances. Thus the ability to select from menus (including the use of icons), to issue natural language commands and to use extremely terse parameterised commands are all favoured under different conditions. There is no reason whatever to suspect that users of mapping systems – as opposed to those of fully-fledged GIS – have requirements which differ substantially.

Perhaps most important is the realisation that there are in many cases two distinct types of user: the experienced and the novice. The distinction can apply in terms either of computer comprehension or of a knowledge of cartographic practice. Since it is the novice users who will be most vulnerable to errors and require most teaching, it is essential to provide a simple dialogue which assists map production. Required therefore is the ability of the package to monitor commands, suggest alternatives where errors are being made and prompt for extra parameters. For the unskilled, menus provide a means of showing all available commands at any one point in the design process.

In addition, the availability of on-line help facilities is now mandatory. These should not only provide definitions and descriptions of all commands and parameters but should also include, wherever appropriate, graphic and text examples. To facilitate understanding, a hierarchical organisation of the help facilities is essential, emphasising the need for a strong and evidently logical conceptual framework for the mapping system as a whole. This help facility needs to be complemented by a compatible set of manuals, including a primer, a user manual, a reference manual and a programmer's manual.

Finally, the increasingly large number of individuals who are producing maps by computer without suitable professional training suggests that the ideal system should also include a demonstrator or even a tutor (perhaps linked to a computer-controlled video tape or disk) to inculcate good cartographic practices. We have been very struck by the preference for good teaching materials available anonymously via a machine, in contrast to traditional forms of supervised 'face-to-face' tuition. For this reason, details of cartographic principles included in manuals may only form a secondary reference source.

Package 'intelligence'

Numerous examples exist of cartographic 'rules' which are widely flouted, often in ignorance. They include the plotting of maps at larger than compilation scale, the production of choropleth maps of areas of varying size based on counts which are not area-based and contouring of non-continuous data. Assuming that descriptions of the data are stored, it is relatively simple to build in such 'rules' as the 'default'. Yet few existing mapping systems provide even elementary safeguards.

Some allocations of symbolism should be programmable as rules at 'run-time'. For instance, the thickening of every 250 m contour might be expressed as:

IF label = 35 AND MOD(label2,250) = 0 THEN LINESTYLE = 2 ELSE LINESTYLE = 1

where label and label2 are attributes of the line indicating respectively the type of line (contour=34) and its altitude above sea level. Equally, all coastlines might be plotted by default in blue, woodland areas in green and railways shown by a line symbol with some element of verisimilitude though (naturally) the user should be able to implement changes where necessary. For choropleth shading, a number of colour sequences can be included – for example, red for population data and green for vegetation. To implement such 'rules' requires that the characteristics of the data be set up in a dictionary which is probably invisible to the normal user.

A further application for package 'intelligence' has already been mentioned. 'Intelligent' user interfaces become more necessary as human assistance becomes less available and the system itself has to act as supervisor. The low standard of much cartography being produced by automated means amply illustrates the necessity for such a facility. To obviate such problems, the user should be told (for example) when there is insufficient space for an outline at a particular place, when a scale bar or key has been omitted and when the output is not (on some acceptable criterion) suitably arranged.

Command files, commenting and the logging of user actions

Many organisations will make repeated use of certain facilities of the system, using the same specifications (notably in agencies creating map series). Equally, many users may wish to repeat virtually all of their previous actions, changing only one or two parameters. For both of these reasons, it is essential to provide a macro-language facility in which commands to the mapping system can be stored, together with suitable comments, in a file for subsequent re-use or editing. The use of command files in teaching is also very helpful. Such files can be created by the user or automatically every time interactive design is carried out. In addition, automatic logging of all commands facilitates the analysis of errors and of system diagnostics when the package fails. Finally, though it is not yet a major problem, such log files can aid the detection of deliberate attempts to damage the cartographic database since they can act as an audit trail.

Response rates
These will evidently be dependent upon the power of the computer, the number of competing processes, the efficiency of the software and the nature of the task requested (including access to any remote databases). In a suitably networked environment, the task should be executed automatically on the most appropriate hardware, but, in any case, a prediction of how long any non-trivial task will take should be provided to the user. Since users require and expect different response rates for different tasks (e.g: interactive editing necessitates a much faster response than is needed for final plotting), the mapping system should also re-assign priorities dynamically to the computer operating system as the user proceeds through the various tasks in a particular terminal session. Processes which are better run in batch mode should therefore receive minimal interactive resources in order to encourage better use of the entire system by the user.

It is our experience that users feel extremely concerned if simple requests, such as a selection from a menu, are not achieved in less than about ten seconds. Therefore, it may prove sensible to refuse to accept additional users in a multi-user environment if the response rates would thereby fall below acceptable levels for existing users – but this must be judged in relation to the type of tasks being carried out, rather than simply the number of simultaneous users.

Independence of hardware and system software
The benefits of this are obvious: the user is not bound to any one supplier of equipment but can choose that which has the best price/performance characteristics and can replace the equipment from different suppliers

if that is judged most cost-effective. This is perhaps partly met by standardising on use of the Unix operating system, except in so far as there are currently different versions of the 'standard'.

In practice, this aspect of the 'ideal' is probably not totally realistic at present because it necessitates that all the facilities available on the largest configuration are also available on the smallest. In addition, the use of an international graphics standard (such as the Graphics Kernal System (GKS)) presently ensures very large overheads in machine processing over a more direct use of device drivers. The safest practical alternative is to assume that all terminals permit Tektronix emulation.

Such a high degree of independence of the host computer system may also be undesirable from other points of view. It is, for instance, often extremely helpful to be able to execute system commands from within the mapping system. This permits information on the system load, the progress of batch jobs and details of directories and files to be obtainable without closing down the mapping system.

Another complication arises because it is unreasonable to assume that all linked hardware in any one organisation will be entirely compatible, at least until Open Systems Interconnection (OSI) becomes generally available. In the circumstances, where mapping systems are not free-standing but are linked to remote databases and the mapping software is resident on a networked computer, then each user needs to invoke a 'login' file which describes the local hardware's characteristics to the mapping programme. How best to meet the user's request for ten colours when only, say, four are available on that particular terminal needs to be defined and set up as a system rule. On a more trivial matter, there is also a need to be able to accept position measurements (e.g: for placing text) in both metric and imperial systems if the package is destined for a truly international market.

Finally, there is one other difficulty which is more severe in cartography than with other computer graphics. This is that almost all present map displays produced on CRTs or plasma displays do not have comparable resolution and size characteristics to those of printed maps. As a consequence, final 'hard copies' produced by high quality electromechanical plotters, often via colour separations for lithographic printing, can be very difficult to emulate on a screen. Thus, on most counts, hardware independence is desirable but is presently impossible to arrange in all respects.

Data independence
At present, the bulk of high quality cartographic systems are vector-

based but image analysis systems are raster-based. Since data in both these structures will inevitably be made available for mapping, and since some operations are best achieved in one structure rather than the other, it is essential that any ideal system should – at least so far as the command language is concerned, if not the graphics results – make the data structure transparent to the user. This may well necessitate the ability to alter data from one structure to another, probably unknown to all but the expert user.

Internal cartographic metafile
This refers first to the need for a common internal format file of the cartographic – and not the graphic (as in GKS) – features. This metafile can be readily interpreted for plotting on different machines. Selected symbolism may also be changed without the need to repeat the whole map generation process. Thus any map or any graphic would be stored as a set of cartographic entities within an appropriate co-ordinate system. Combined with a panelling capability, this would enable the user to overlay separately produced images, or alternatively, to generate automatically sets of images in a layout suitable for direct transfer to the printing plate.

Statistical summaries of the data
It is frequently of great value to know how much data exists, e.g. after various selection operations have been carried out. This facilitates design decisions and also permits prediction of plotting time and cost. The descriptive statistics required are basic ones, notably the number of features in each class, total line length, and the mean, minimum and maximum polygon size at plot scale.

Creation of new mapping capabilities
It is impossible to predict all the mapping requirements which will arise in the future. For this reason, the ideal mapping system must also be extensible. Two methods of achieving this seem obvious. The first is to permit the creation of new commands by generating macros from pre-existing primitive commands as described earlier and enabling these new commands to be stored in and documented within the system. Thus particularly complex symbols could be built up from simple ones already held in the system. The second (and complementary) approach is to provide a programming interface so that skilled users could, where absolutely necessary, add to the existing code to produce new functional capabilities.

Map Design and Construction Capabilities
In general, these include the capabilities set out below.

Different types of mapping
The minimum range of types of mapping which should be available consists of the following:

—point symbol mapping
—line symbol mapping
—'area-fill' mapping
—text mapping
—surface generation and depiction
—summary statistical displays
—any combination of these

The first three of the above need to be able to operate appropriately on both qualitative and quantitative data. Thus a single symbol may be used to represent one class of road, but the same basic symbol may also be used to display flows of traffic along the road if one of its characteristics (e.g: width) is allowed to vary in response to the value of this one attribute.

Text mapping is sensibly regarded as a separate type of computer mapping for two reasons. The first is that it frequently accounts for much of the time and cost of generating maps and the second is because of its considerable complexity – compounded in systems which cannot generate publication quality lettering on a terminal screen. This complexity chiefly arises from overlap of text strings with other text strings or with other types of important map features, leading to the need for 'shielding' and other such graphic devices. So many choices exist in name placement that a number of heuristic based 'expert systems' have already been created to attempt to automate the process, with varying levels of success.

The ability to interpolate a surface from haphazardly distributed XYZ observations (such as of temperature or altitude) or from the equivalents of XYZ values arranged in a linear structure (e.g. as contours or along a ship's track) is a necessary preliminary to the display of that surface in map or oblique view form. The latter includes perspective and orthogonal views, inclined contour models and hill-shaded relief. It should be noted, however, that the generation of the surface may be achieved using a multiplicity of algorithms and the choice of method can only be made sensibly with expert knowledge of the characteristics of the data and some understanding of the phenomena being modelled. To that extent, successful surface modelling often requires collaboration between (for example) a geologist and the cartographer.

It is now a matter of routine that analysis of geographically distributed data normally begins with simple statistical descriptions. For this reason and to summarise the non-spatial aspects of the data, it is essential that the mapping package can compute the necessary statistical tests and portray such basic diagrams as histograms, pie graphs and line graphs.

Finally, it must be possible to combine any of these types of mapping if necessary. The finished result in any instance must be capable of being interactively edited to obviate unacceptable overlaps of features. In order to enhance clarity and legibility, chosen symbolism needs to be applicable to any desired class of cartographic feature. A range of default symbolism has, of course, to be provided and amply illustrated in the manuals. We have already pointed out that default symbolism should exploit verisimilitude wherever possible but, since these defaults may not always be suitable for the user's express purpose, an interactive symbol editor must be included. Menus can be implemented as one means of assisting inexperienced users.

Generation and placement of conventional cartographic elements
These include the following:

—Boxes/frames which may be generated (e.g. as a rectangular box, as a circle or as a section of a graticule) or defined interactively. Like all other conventional elements, this should be allocated an attribute and be treated henceforth as part of the map and be capable of being edited. It should therefore appear in the cartographic metafile.
—Scale bars, generated automatically from the system's knowledge of the map units and extent.
—Orientation indicators, such as a north arrow. Again, this should be computable from information available to the system.
—Keys and legends. These will vary greatly depending on the type of map, the number of classes of feature to be shown on the key, the shape of the area for which data are available compared to the plotting frame and personal preference. It is possible, however, to generate default keys and legends and, provided the metafile is hierarchically structured, both to 'tow' these into desired positions on the map and to re-scale them appropriately.

All these components should appear automatically (if appropriate) and only be omitted after repeated questioning of the map designer.

Transformations of data
These need to be available for both the geometric and the attribute elements of the cartographic data. In the former case, they need both to enable conversion from representation in one co-ordinate system to its equivalent in any other algebraically expressible one through a change

of map projection and also to correct local distortions through use of control points at known positions ('rubber sheeting'). Transformations of attributes which are required include recoding (e.g. to form classes) and, where the attributes are measured on a suitable scale, 'normalisation' and skewness-minimising transformations should be available.

Selection of data
Again, this must be possible in both the geometric and attribute 'domains'. Such 'windowing' needs to be supplemented by the ability to operate on multiple attributes using arithmetic and Boolean logic. Moreover, this must be possible without restriction on the numbers of attributes considered and must be possible at any desired stage in the mapping package.

Generalisation of data sets
The need to simplify or generalise data sets is well known to all cartographers. The shortcomings of most existing automated cartographic generalisation tools are, however, little appreciated. The fundamental mismatch between cartographic 'entities' or 'objects' (often selected at the digitising stage for operational convenience) and 'real world objects' has ensured that the basis on which the generalisation occurs is an inappropriate one to mimic human generalisation or to work under reasonable 'rules'. To achieve satisfactory generalisation, then, it is necessary to work on 'real world objects' such as a road pattern or a surface, rather than upon part of their cartographic representations (e.g. one side of the road or one contour at a time). Context must also be taken into account. All this ensures that only certain data structures can facilitate the generalisation operation.

Output Characteristics
These are readily stated in qualitative terms:

- —The ability to produce output on vector or raster plotters irrespective of how the data are stored in the system and held in the cartographic metafile.
- —The ability to sort the metafile to ensure minimal time for plotting.
- —The ability to so arrange the default symbolism as to cope with terminals of much less than ideal capabilities (e.g. those with very few colour capabilities).
- —To be able to produce colour separations automatically if required.
- —The ability to produce identical output in both hard and soft copy.
- —The ability to produce output at any scale and also to scale a metafile prior to plotting.

CONCLUSIONS

Several conclusions may be drawn. The most important of these are set out below:

—A significant number of the characteristics of this ideal mapping system are not easy to meet at the present time. Such difficulties are mostly those arising from deficiencies in either the hardware or the system software. This is often exacerbated by local shortcomings: for instance, it is sometimes very difficult to emulate high quality line screened output readily shown on a high resolution raster screen on a pen plotter.

—Though the mechanical aspects of cartography can often now be replicated very satisfactorily, those aspects which have traditionally been achieved by human judgement – notably generalisation and name placement – are still poorly tackled by existing operational systems.

—To remedy the problems identified requires a more sophisticated approach to data organisation than is required simply to plot graphics. It may well, for instance, necessitate adoption of topological data structures and 'object programming'.

—At the same time, their solution requires a greater interaction between cartographers and computer system designers to rectify the omission of basic cartographic expertise embedded in mapping packages.

—The growth of cartographic data libraries is inevitable since no one organisation is capable of encoding all desired geographic data. Many of these will constitute large data files, accessible over telecommunications links and requiring major investment to maintain and up-date. The 'ideal' package will therefore need to be able to access these as a matter of routine, without informing the user of its actions other than by providing an estimate of cost before taking irrevocable action!

—No firm indication can be given of the 'eventual' form and structure of the ideal system. One likely path, however, is towards GIS with the mapping package as one tightly integrated part of this tool kit.

ACKNOWLEDGEMENTS

This paper is based upon an idea originally suggested by Jack Dangermond. Thanks are also due to Teresa Connolly for helpful comments on drafts.

REFERENCES

Keates, J.S (1973) *Cartographic Design and Production.* London: Longman.
Robinson, A.H., Sale, R.D., Morrison, J.L. and Muehrke, P.(1984) *Elements of Cartography*, fifth edition. New York: John Wiley & Sons.
Waugh, T.C. and Chulvik, C. (1982) 'State of the Art' Graphics Presentation: GIMMS release 4, in Rhind, D.W. and Adams, T.A. (eds.) *Computer Cartography*, British Cartographic Society Special Publication no.2, London (eds.).

THE REVOLUTION IN CARTOGRAPHY IN THE 1980s

Joel Morrison

INTRODUCTION

Cartography is in a state of near-revolution. Never before in the history of cartographic innovations has the discipline experienced such drastic technological advances in data capture methods at the same time as it is experiencing dramatic advances in its production technology. In the past, new data sets have often been created by changes in technology, such as from the increase in accuracy of land measurement traced to mathematical advances made by Arab cartographers of the pre-European renaissance period, the increased accuracy of measurement at sea resulting from the invention of the chronometer and the introduction of the systematic censuses of the late 18th century. At other times, new technologies for cartographic production have been developed following such innovations as the invention of printing, of photography or of plastic materials. Today, however, cartographers are simultaneously experiencing drastic changes in both data sources and production technology. This makes the current period unique in the history of cartography.

This uniqueness is also characterised by the increasingly complex set of procedures and processes which are in use in cartography today. When coupled with broad generalisations, these may lead to false understandings of contemporary cartography. For example, the fact that technologically advanced countries – the so-called first world nations – are leaders in the technological developments, coupled with an assumption that they bear a social responsibility to aid the developing nations of the second and third worlds, could lead one to the mistaken and simple expectation of a 'trickle-down' of data capture and production technology from the first world to the second and third world nations. Central to this concept is the premise that a large gap exists between cartography as practised in the technologically advanced nations and cartography as practised in the rest of the world. This is simply not true. Today, cartographers share the rapid advances in production technology in first, second and third world nations.

Since 1980, this author has travelled extensively as President or as a Vice-President of the International Cartographic Association and I have been impressed with the fascination of all cartographers with the new production technology and the new sources of data. In the middle of China I found a state-of-the-art, computer-driven image analysis system. From India I saw 1:50,000 scale maps being used to create a 1:100,000 scale map by computer-driven technology. I acted as a consultant on the use of a micro-computer system for wildlife management in Rwanda. At the same time, I talked with some senior cartographers within the United States, the United Kingdom and France who had no plans for utilising the new computer technology in their cartographic production. They were unaware of developments and unable to tap the resources necessary to enter today's arena of computer-driven cartography. Clearly, the need to disseminate knowledge of and transfer new cartographic technology is essential but that transfer is not only from developed nations to developing nations; the need also exists within all nations – particularly within the most technologically advanced ones.

This new production technology represents an additional technology which all cartographers can use with benefit. There is, of course, no need to abandon the older cartographic technologies and in fact it is often an advantage to possess multiple technologies to produce map products. Our ability and need to use the new technologies is, however, an additional benefit for mankind. As the internationalisation of society increases and as problems become more and more global in nature, so the task of communicating spatially is growing more complex, and as the need to communicate spatially is growing, the potential role of maps grows ever larger. Fortunately, our capabilities to utilise the new technology can allow that communication to take place more rapidly. Cartographers have the means, therefore, to meet important communication needs of the world society of the late 20th century.

What, then, are the major problems confronting the cartographic discipline in the 1980s? What trends are evident? What is our future?

To present a sketch of the state of cartography in the 1980s from one perspective, this paper is divided into four parts. First, the need for more rigorous definitions in our field is acknowledged and discussed. Next, the various stages of the cartographic process (data capture, data manipulation and data visualisation and generation of products) are analysed in relation the changes which are occurring. Third, the organisation of current world cartographic activity is briefly characterised. Finally, a brief outline is given of the significance of the identified trends for the future of cartography.

THE NEED FOR DEFINITIONS AND REDEFINITIONS IN CARTOGRAPHY

The revolutions in technology for data capture and map production in cartography are causing several traditional disciplines with close ties to cartography to reassess their role and content. Nowhere is this more evident than in the field of photogrammetry. By exploiting the full potential of the computer, it may be possible to by-pass the manual photogrammetric compilation stage of traditional cartography. In several places around the world – China, India and South Africa – where photogrammetrists themselves have recognised this, we see an increase in the use of close-range photogrammetry to aid other disciplines and restoration projects. It may well be that photogrammetry as we have known it will be less utilised in cartography in the future than in the medical sciences and archaeology. On the other hand, some photogrammetrists are trying to develop their relevance to topographic science and are fighting hard to re-define themselves – and at the same time cartography – so that they will still remain an important stage in the science of mapping. It will be interesting to watch this scenario play out in the years ahead. Surveyors are likewise affected by changes in technology. With increasingly general use of the Global Positioning System (GPS), the practice of land surveying will be radically changed. Many surveyors are now turning their attentions to Land Information Systems (LIS) as a result (CIS 1985).

The nature of the technological shift taking place in mapping affects not only our colleagues in traditionally neighbouring disciplines but also must affect the definitions of processes and procedures which cartographers use. In general, the definitions which we use in the manual and traditional production technologies are not specific or rigorous enough for use in the arenas of computer or electronic technology-based production. An obvious example is the definition of cartographic generalisation. The textbook examples of generalisation are generally not programmable without a greater degree of specification (see, for instance, Robinson *et al* 1984). Cartographers need to seek collective agreement on these specifications.

Production processes are also affected by the possibilities of increased precision made feasible by the use of computers and high resolution plotters. For example, the difference between basing a map projection transformation on an ellipsoid as opposed to a spheroid was in the past often not considered worth the added manual labour required to perform the more involved calculations. Moreover, it was frequently regarded as nearly impossible to draft the improvements in results of the calculations by manual means. With the computer doing the calculations and a high-resolution plotter drafting the results, the more accurate results from using an ellipsoid can now be achieved as easily as the less accurate results based on a spheroid. Therefore, the criteria used by the cartographer in taking a decision on the particular procedures to be used in making a map are also changed. It is, of course, necessary for the cartographer to be trained appropriately so that he or she can make informed decisions on when the increased accuracy is actually needed.

Given all this, cartographers should not be surprised when we find that our field and its processes and procedures need redefinition. As has been pointed out in other writings, the official definition of cartography by the International Cartographic Association does not mention computer-assisted cartography, nor the new products that result from it (ICA 1973). It is vital to our own self-interests that cartographers redefine the discipline for the future in the context of all of the technologies which are now available for use.

Not only is it necessary to reassess the definitions of our areas of expertise; we must also specify our standards. Cartographers have always paid close attention to accuracy and precision in map-making. And to a great extent those terms have been defined on the basis of the technologies in use at any particular time. We need to re-define our standards in terms of the new electronic technology, or at least confirm that existing standards remain relevant when using that new technology.

The United Nations model of cartography, as used in the UN Regional Cartographic Conferences, could serve well for the purposes of examining the definitions of our field (UN 1987). The UN recognises four sets of processes: data capture, data processing, data visualisation and the management of cartographic activities. We need to consider these sets of processes systematically and make the necessary changes to definitions. The new definitions must be useful for all technologies in use in cartography today.

In the discussion that follows, the UN model of cartography will be used to focus attention on each of the four sets of processes which that model specifies. Through this mechanism, we will seek to understand how existing and new trends will influence our future.

Data Capture

Clearly there have been major advances in data capture in the last thirty years and advances are continuing to be made. We now have satellite-captured electronic images as well as satellite-derived photographic images. (Most sensors can of course be used both from aircraft and from satellites.) We have broadened that portion of the electromagnetic spectrum from which we can record data and transform them into image form. We have also increased the spatial resolution with which we can collect these data.

Nothing illustrates the last statement as well as the trend that is openly documented in satellite remote sensing as carried out within the visible spectrum and the near infra-red region. Starting with the Landsat series of earth resources satellites in 1972, we see an increase in the spatial resolution of the data that can be captured from space for civilian use. Early Landsat data had a resolution of 80 m. In theory, this meant that earth features which were at least 80 m in one dimension could be located from space. The major scientific significance of the Landsat series of sensors was due to the 'open skies' policy adopted by the United States government about the distribution of the Landsat images. Never before had systematic and repeatable world-wide coverage at 80 m resolution been readily accessible to every citizen of the world. Beginning with Landsat 4, a sensor package called 'Thematic Mapper' was included, and the resolution of the available data was improved to 30 m. Again, the 'open skies' policy was in effect. National Mapping Agencies quickly determined that 30 m resolution was sufficient to do mapping at scales as large as 1:100,000 (Calvocoresses 1986).

In 1986, France launched the SPOT satellite. This made available for the first time digital data from space with 10 m resolution (CNES 1984). These 10 m data are collected in panchromatic form, using a Multiple Linear Array solid state sensor. The SPOT satellite also includes a colour sensor with a resolution of 20 m. The 'open skies' policy is still in effect to those with sufficient money to purchase the imagery. Most recently, the USSR has announced that it will sell to the world photographic data with 6 m resolution (SK 1987).

As the reader can see, the resolution of space imagery has shown dramatic improvements within the past 15 years. Parallel to these developments has been the expansion of the capture of data in other regions of the electro-magnetic spectrum. Imagery of the earth resulting from sensors in the thermal infra-red regions and microwave and radar sensors are all proving their usefulness. One of the next generation of NASA sensors will record data in no less than 100 wavebands. It may well be that, as such improvements continue, we will be able to specify different sensors for each data set that we wish to capture. Cartographers may no longer have to rely on a single image reconstructed on earth from impulses received from a space-borne sensor to yield data about multiple geographical distributions. In effect, we may be able to specify the sensor best-suited for hydrography, a different one for vegetation, one for ridge lines, one for transportation features and so on.

The improvements obtained in data capture from satellites are also being exploited to improve our abilities to capture data from aircraft. To cite but one example, the U.S.G.S. has developed a sensor that very accurately measures the variation in the terrain (Chapman and Cyran 1988). The accuracy appears to be within fractions of a foot and the sensor would even appear to be useful in regions of very flat terrain. Likewise, we have continued to make improvements in the films and filters that are used on airborne cameras.

The Global Positioning System mentioned earlier allows the more accurate positioning of points on the surface of the earth (Wells *et al* 1986). When the GPS system is fully in place, it will reduce the amount of technical knowledge and the number of calculations that a surveyor who works in the field will need to know and make.

There remain problems in the area of data capture. A major problem is that of the high cost of the equipment that is required. Clearly, not everyone can afford to launch a satellite for data capture purposes. A continuing problem, however, even for those who can afford the data capture equipment, is the equally severe one of the volume of data that space-borne or airborne electronic sensors are capable of returning to ground receiving stations. These are sometimes orders of magnitude greater than the volumes of digital map data, leading to real problems in processing, storing and distributing the mass of these data to potential users.

The net result of this revolution in data collection is that cartographers must deal with enormously larger volumes of data when in an electronic environment (Light 1986). These data are transmitted at high speed and can be manipulated to advantage only if they are properly managed. In many cases – and in contrast to the traditional problem – the cartographer now has too much data for a map. In these circumstances, this transforms the problems of map inaccuracies from ones of insufficient data to ones arising from the use of inappropriate methods of processing of the available data. In summary, a cartographer must now be knowledgeable on how to utilise the masses of existing data throughout the whole cartographic and data manipulation process to achieve a map's purpose.

The problem from the perspective of the world body of cartographers is even more complicated. If there exists a surplus of data from those who have the capabilities to launch satellites and to receive their signals, there also remains a large number of cartographers who do not have ready access to these data (for either economic or political reasons). The ability to utilise these data has an educational as well as an economic dimension. A third world country may have the capital to buy the data – and perhaps even the equipment with which to process the data – but the country must also have the technical expertise amongst its workforce to carry out the processing in a way which minimises the chance of misleading inferences being drawn from the maps or imagery. Sometimes the short-term cost of educating the specialists is greater than the actual purchase of the equipment and data, and it certainly takes longer to educate than to purchase. A continuing need in cartography for the transfer of the technological knowledge exists. ICA has already helped in this regard and can continue to do so.

Data Manipulation
Cartographers have always needed to manipulate data to produce their maps. A keen sense of how to perform these manipulations is recognised as one characteristic of an expert cartographer. However, today's events require that we make a science out of the art of cartography. Cartographers are being presented with a wide array of manipulation tools. Some of these tools are born out of necessity, and some are simply available as a result of the benefits of the electronic technology. It is convenient to discuss these data manipulation activities under two categories: first is data structuring and maintenance and second, statistical processing of data and processing of the data for visual display.

Given the vast amounts of data involved, it is mandatory that the storage and manipulation of that data be accomplished in such a way that the results can be used efficiently. For example, one can easily see the advantage of knowing that a boundary is coincident with a river. In most computers, it is usually more efficient to have one encoded data string to represent both features, the boundary and the river, than to maintain two identical data strings. Likewise, it is often important to know the point at which two lines intersect. To accommodate this information in the data, some element of structure needs to be added to a data set. It is not surprising, then, that cartographers are working increasingly with computer scientists in researching new ways in which to structure and store cartographic data.

Primarily, data structuring and maintenance consist of creating and editing computer-compatible files of data. In doing this, a cartographer must be able to reformat existing files, define (or compute) topological relationships between elements of data and add these to a file, derive new data files from existing files and integrate, revise, validate, and correct files (Morrison 1987). As cartographers become more expert at data structuring, numerous large multi-purpose data sets will become available for use in making new cartographic products.

The capabilities for statistical processing of cartographic data have been greatly expanded by the introduction of electronic technology. The traditional transformations required for projection changes or co-ordinate system calculations can now be done more efficiently and much more quickly. As pointed out earlier, some projection transformations of the earth onto a flat surface were simply avoided in the past because of the complexity of the required calculations prior to the availability of computers. An entirely new class of transformations, those who meet some stated statistical criterion, is now available. One such example is Snyder's least error projections for Alaska (Snyder 1986).

Scale changes, data sampling, statistical filtering and smoothing, classification, interpolation and the use of map overlay techniques can all be accomplished more efficiently using the electronic technology – provided, of course, that the necessary cartographic data and software are available. The capability to have dynamic displays rather than static maps allows increasing use of time-sequenced data. Work on all these enhanced capabilities for cartographers is on-going but time-consuming. Finally, such research is also being carried out into ways of making machines more 'intelligent', thereby obtaining (in principle) both the human problem-solving capability and the error-free nature of computation offered by the computer.

Data Visualisation and Cartographic Products

Today cartographers are no longer limited to producing one type of product. A map consisting of symbols on a sheet of paper is only one of many options that cartographers have at their disposal. This wide array of new products includes not only printed maps but also digital files, temporary displays on CRTs and interactive map construction and use. As a result, a new vocabulary of terms – virtual map, pan, zoom, refresh, CD-ROM and much else – has entered the traditional cartographic lexicon. Philosophically, this change in products is important for cartography and at this time its effect is probably under-estimated.

Cartographers are purveyors of spatial data. Prior to having those data available in computer-readable form, cartographers controlled what was done with their data. Procedures (agreed often only as 'rules-of-thumb') and conventions enabled the cartographer to make a map which, it was believed, communicated well. Often the communication transcended different languages insofar as an English-speaking human could read without difficulty information from a large-scale Japanese topographic map, except for the place names, and vice versa. Today, many cartographically untrained users can access digital data. To the extent that the users do this, the cartographer no longer controls the use of cartographic data. Cartographic conventions and rules-of-thumb are unknown to many potential users of spatial data. North orientation, blue water, equivalency for showing densities, darker values for more values – these and all the other conventions do not need to be followed by naive users. Either humanity learns to use the available cartographic data through trial and error, or cartographers must teach them how to utilise their data effectively. In either event, the control has irreversibly shifted from the cartographer to the user.

Moreover, new distribution media make cartographic data and products available in strikingly different forms. No longer is a map limited to being a systematic placement of symbols on a sheet of paper or on a globe representing a scaled reduction of the earth. Maps can be stored on, transferred by and reconstituted from video disk; they may also reside in several forms of storage that computer terminals can access. Cartographic data now exist on tapes, CD-ROMS, optical disks, etc. It makes it very difficult even to define a cartographic product.

Technologically, it is now perfectly feasible to view data electronically on a screen which represents a map containing the information which you have requested. The BBC's Domesday system initiated the concept of a 'surrogate walk' in eight parts of Britain but it is only a matter of time until data are available and accessible which will allow you to experience a walk or drive down *any* street in the world. A moving display literally could bring the world to you, adding a new dimension of realism to the term 'armchair geographer'. Imagine the visual ability to 'walk down' the Champs Elysées or being in Tiananmen Square, being able to turn left or right or to pivot and to view the scene as you move through or along a path that you control – all from the comfort and convenience of your armchair or office. Other industries will be able to supply the auditory and olfactory senses with appropriate stimuli, but cartographers must be prepared to stimulate your vision.

Developments in hardware have been spectacular and are continuing at a rapid pace. Further spectacular developments in supercomputers, parallel processors, storage capabilities, local and wide area networks, narrow and wide band communication channels, fibre optics, and improved satellite transmissions seem assured.

Software development is moving at a slower pace and is probably the area most in need of cartographers' attentions today. Just as vital is the need to develop 'rules' for the automated extraction of cartographic data from remotely-sensed data. At the other end of our processes, we need to develop production 'rules' for the automatic production of our products. Automatic lettering placement on cartographic products is illustrative of the software problems facing cartographers. Precedence in automated cartographic symbol placement is another area in need of further attention. Cartographers should be part of these exciting developments.

One other important trend is to make a team of a professional cartographer working with a computer programmer. This is the result of past attempts to have trained cartographers do computer programming and also of attempting to have trained programmers substitute as cartographers. Neither proved to be efficient. Apparently, the best combination is for a cartographer who has some knowledge of programming and computer systems to have at his disposal access to a well-trained computer programmer.

CARTOGRAPHIC ORGANISATIONS

Much cartographic activity is performed by national, regional and local governmental organisations in almost all countries of the world. Cartography, for the most part, is an activity of government. Certainly it should be so because governments have a vested interest in their own territory. Maps of the territorial extent of any nation are a prerequisite for national protection and for planning and development of any kind. Private cartographic enterprises exist in some countries but they tend to concern themselves with educational needs for maps and with maps needed by individual citizens or for specific projects.

Governmental cartography is done on a national scale by military establishments in most countries. Some countries also have a strong tradition in civilian mapping by governmental agencies. For the most part, however, governmental mapping is general-purpose mapping at relatively large scales. Governments often will create what is essentially a base map, upon which other distributions are plotted by other users to aid in planning and development. These base maps may be topographic in nature. The same activity may also be carried out by the military, but the objectives to be met are somewhat different.

In many countries today, basic mapping remains closely guarded and the free distribution of government-produced maps is not allowed. In other countries, large-scale topographic maps are made freely available to the citizens and anyone else who wishes to procure them. It will be interesting to see how the recent developments in high-resolution data capture by space sensors will affect those nations which still do not allow the free circulation of their basic topographic maps.

In viewing the world organisation of cartographic activity, time must be allocated – even in the 1980s – to acknowledge the historical importance of the British Empire in the field of cartography. One important activity that the British colonials did accomplish, which has proven to be advantageous to all those nations formerly part of the British Empire, was the export of cartographic expertise. Most nations which were at one time or another (particularly in the 19th century) under British influence are, relatively speaking, well-mapped today. The British placed a high priority on mapping. To them, mapping was important and essential to the domination and administration of a territory. Some of those strong ideals remain in many nations – even after independence. India, South Africa, Kenya, Nigeria, Canada, Australia and, of course, the United Kingdom itself, have long histories of excellent large-scale topographic mapping.

In contrast to the British influence is the colonial heritage of many of the Latin American nations. In Latin America, it is more common to have military mapping organisations than it is to have civilian governmental units in charge of large-scale general purpose mapping. While not ubiquitous, in most cases and as far as the average citizen is concerned, those countries are relatively poorly mapped.

Along with this difference is an enhanced general appreciation for maps amongst the citizenry of those nations with a long history of mapping. On the other hand, the current availability of satellite imagery may tend to make accurate maps of all the world's territories more readily available for peaceful uses. This may well enhance the cartographic awareness of the entire world's population.

Because mapping is so important to national governments, it is a fact that in the world today the most advanced systems for the production of maps are often run by governmental units. The entry costs to this mapping are high. In many instances, the costs can only be afforded by the government because the government is the only entity that has a fundamental and demonstrated need for complete cartographic coverage of its territory. The activity itself cannot be expected to be self-supporting on a commercial basis. Furthermore, satellites with sensor systems of utility for cartographic purposes are currently owned by governments. France has designed and launched the SPOT satellite with a combination of private, quasi-private and government support and other planned satellites will probably be financed in the same fashion. But all other existing satellites for mapping purposes were at least initially government-owned. For the most part, therefore, if you wish to see the most advanced cartographic technology that exists in a given country, you must visit a governmental agency in that country.

Private cartographic activity in the world is limited to those countries which allow a free-enterprise system to operate. Some quasi-governmental organisations that sell maps in some countries operate somewhat like private cartographic enterprises which exist under government mandate. However, it is important to remember that, in commercial terms, it cannot be cost-effective to do large-scale coverage of an entire nation. One therefore finds private cartographic enterprises limiting their involvement in national coverage to doing atlases at relatively small-scales. If a private agency does map at large scale, it is usually for a well-defined purpose over a relatively small areal extent and for which there are many demands. For example, a private enterprise may well prepare a street map of a major city, a national park tourist map, or a map for educational purposes. The design of the maps produced in each one of these mapping projects may well differ significantly.

Due to the profit requirement, for the most part private cartography in the world lags behind the governmental mapping agencies in the employment of the latest electronic technology. There are many good reasons and some potential benefits for this, but the primary one is that the conversion to electronic technology – and, in particular, the conversion of data into machine-readable form – requires a tremendous economic outlay on hardware and software. Moreover, hardware and software are still in rapid phases of development and so it is very risky for private groups to spend the capital necessary to purchase each new generation of hardware and software.

Another reason why private cartography lags behind governmental cartography in most nations is that often private cartography (the so-called value-added industry) utilises the results of governmental mapping to form the base for their products. Therefore it is necessary for the government to make a 'core amount' of cartographic data and products available before private cartography finds it profitable to move over to use of the electronic technology which will allow it to utilise the government's data and produce new products.

Academic cartography has also been affected by the introduction of electronic technology into cartography. Academics cannot usually afford capital equipment in the form of hardware and software, except when provided by government. To date, textbooks for university courses which detail the latest advances are not readily available. And perhaps most important is the fact that entrenched faculty in many universities are not trained in how to exploit electronic technology. It is almost essential that faculty members be retrained if we expect them adequately to train our most promising youth in the latest advances in cartography. This is not to say that it is impossible to educate a new breed of cartographers in academia. Certainly some education can be done without the latest equipment and books. As was mentioned earlier, the philosophical or theoretical aspects of cartography have not changed so radically. There remains a generic way of doing cartography, and this has been transferred into the electronic production environment. The steps in the cartographic process have not changed as much as the manner and machines on which they are performed. Therefore, it should be possible to educate students in the theory of cartography, regardless of the technology available. But the governmental agencies – and more recently the major private firms – need to be able to hire well-educated and trained individuals who know both the theory and the current technology of cartography.

It is sad, therefore, that our academic institutions are in such a state. Today in the United States, and I am told in other nations as well (notably the UK), academic salaries and equipment grants have not kept pace with salaries in government service or in the private sector. Therefore we have Professors who hold positions in our Institutions of Higher Education who are unfamiliar with the electronic technology, do not have the funds to purchase the electronic technology and who have little or no incentive (in the form of remuneration) to do anything about the situation.

Fortunately, there are a few hopeful signs on the world map in the form of newly renovated curricula, newly established specialities and the formation of 'centres of excellence' in education. The concept of a 'centre of excellence' for cartographic education and training in a nation has a certain appeal, and I find this excellent idea being discussed in many places in the world. Cartography is a small discipline and the world does not need many cartographers, yet cartography is now a capital-intensive discipline. The 'centre of excellence' concept allows the clustering of new technology and academics in one physical location, which should generate a synergism. Time alone will provide us with the answer to the wisdom of employing the concept in cartographic education and training throughout the world.

The future also holds out the potential for consortia, composed of governmental agencies and private firms working with academics associated with a centre of excellence, to become the leading mapping innovators in many nations. Japan has tested and used this idea in several industries and France has done somewhat the same with the SPOT satellite organisation. If one can speculate that at some point in time concern for humankind will outweigh concern for territorial limits, then we can hope to see a truly international cartography composed of government, private interests and academia working together to improve man's habitation on this earth by producing the necessary cartographic products.

In many countries, professional organisations of cartographers have formed and journals relating research, development and other cartographic activities have proliferated in the past three decades. One could easily make a case based on these developments that the discipline of cartography, as separate and distinct from geography or engineering, was firmly established in the mid-twentieth century. The amount of research into cartographic topics has certainly increased, but it remains inadequate.

Just as the Second World War spawned a new generation of cartographers, the development and use of electronic technology in cartography is spawning another new generation of them during the 1980s.

The ICA epitomises these developments. The outgrowths of rapidly changing technology both during the Second World War and currently contributes to renewed invigoration of cartography. The ICA was founded in 1959, and in the 1980s a new generation of cartographers has assumed the roles of leadership in the organisation. What, then, is the future of cartography which they will have to face?

THE FUTURE

The trends described in this paper all stem from the fact that cartography has been experiencing a revolution brought about by the coincident occurences of a new technology and a new source of data. The effect has been rather all-encompassing, changing procedures, processes and products. Relatively little change in basic philosophy or theory has, however, occurred. Whether changes in philosophy and theory will occur in the near future is an interesting but open question. The potential is present, but tradition in cartography is strong.

Because of the changes, cartography today requires a new definition, and redefinitions of many of its processes and procedures are also required. Existing standards must be re-examined. Increased data resolution, expanded capabilities for sensing in various parts of the electromagnetic spectrum, and the use of both airborne and space-borne sensors result in a huge increase in the volumes of data that cartographers have available to exploit. Data structuring and new statistical processing techniques are required and make the actual work of the cartographer easier, as machines accomplish the tedium associated with the former processes. Many new products, whose full potential can only be estimated at this time, are now available. New media in which cartographers can work are available. The transfer of these technological developments is needed both from the technologically advanced nations of the world to the developing nations and, within the technologically advanced nations, to all agencies requiring cartographic products.

In spite of all the technological change, only one philosophical change has been noted. As large digital data bases become accessible, the cartographer loses control over the use of that data to a group of (at present) cartographically naive users. Problems in communications will undoubtedly result. Whether the technological revolution will lead to a more profound philosophical revolution, or even to philosophical evolution, remains to be seen.

The revolution of the 1980s in cartography, then, is one of technology, not philosophy. We still need to communicate spatial displays of the patterns existing on the earth. That is still the primary goal of cartography. We continue to need cartographic products for defence, development and planning, as well as for education. Cartography has established itself as a discipline in the twentieth century. National mapping organisations have become the leaders in the new technological changes that are revolutionising cartography. Academic cartography has not kept pace, yet the emergence of centres of excellence or research institutes in cartography promises to allow a synergistic effect to be created between government and academia, which will foster even further advances in cartography.

This is an exciting time throughout the cartographic world. One doubts whether simultaneous large magnitude changes in both data-gathering and production technology have ever before or will ever again take place. Let us each and all enjoy our involvement in our chosen field and look with excitement to the future break-throughs in cartography.

REFERENCES

Chapman, W.H. and Cyran, E.J.G. (1988) Airborne Precision Mapping System, US Geological Survey Open-File Report No. 88–101, 11 pp.

CIS (1984) 'Decision Maker and Land Information Systems'. Papers and Proceedings from the FIG International Symposium, Edmonton, Alberta, Canada, October 15–19, 1984, The Canadian Institute of Surveying, Ottawa, Ontario, Canada, 1985.

CNES (1984) Centre National d'Etudes Spatiales; SPOT – Satellite Based Remote Sensing System, CNES, Centre Spatial de Toulouse, Toulouse, France, March 1984, 15 pp.

Calvocoresses, A.P. (1986) Image Mapping with the Thematic Mapper, *Photogrammetric Engineering and Remote Sensing*, 52, 9, 1499–505.

ICA (1973) *Multilingual Dictionary of Technical Terms in Cartography*, Franz Steiner Verlag GmbH, Wiesbaden, Germany, p.1.

Light, D.L. (1986) Mass Storage Estimates for the Digital Mapping Era, *Photogrammetric Engineering and Remote Sensing*, 52, 3, 419–25.

Morrison, J.L. (1987) *Cartographic Data Manipulation*, Document E/CONF. 78/BP.2, U.N. Economic and Social Council, 11th U.N. Regional Cartographic Conference for Asia and the Pacific, Bangkok, January 5 −16.

Robinson, A.H., Sale, R.D., Morrison, J.L. and Muehrcke, P.C. (1984) *Elements of Cartography*, 5th edition, John Wiley & Sons, New York, N.Y., pp. 124 −132.

SK (1987) Sojuzkarta-Kartex; Topographic-Geodetic, Aerial Survey and Cartographic Activities, Sojuzkarta-Kartex, Moscow, USSR, 7 pp.

Snyder, J.P. (1986) A New Low-Error Map Projection for Alaska, *New Frontiers, ACSM Technical Papers*, American Congress on Surveying and Mapping and American Society for Photogrammetry and Remote Sensing, Fall Convention, Anchorage, Alaska, pp. 307–314.

UN (1987) U.N. Economic and Social Council, Provisional Agenda, Document E/CONF. 78/1, 11th U.N. Regional Cartographic Conference for Asia and the Pacific, Bangkok, January 5–16, 1987.

Wells, D. *et al.* (1986) *Guide to GPS Positioning*, Canadian GPS.

INDEX